Fuzzy Logic 2

How to Keep Thinking Like a Veterinarian

Printed in the United States of America

First Edition, 2022

ISBN 978-1-955639-06-4

Cover design: Patience Parks-Young (patience-creative.com) and Aaron Gruben.

The pictures and photos not drawn by the illustrators and not radiographs/ultrasounds are public domain.

Every effort has been made to attribute correctly all the material reproduced in this book.

Every jot and tittle of the text by Aaron Gruben (except where otherwise noted).

Post Tenebras Lux Books

www.nmgrubens.com

www.posttenebrasluxbooks.com

Fuzzy Logic 2

How to Keep Thinking Like a Veterinarian

Written by J. Aaron Gruben, DVM

Illustrations by J. Aaron Gruben, Geneva Gruben, and Catharina Gruben

Also By J. Aaron Gruben

Veterinary Books

Fuzzy Logic: How to Think Like a Veterinarian

Fuzzy Logic 2: How to Keep Thinking Like a Veterinarian

Fuzzy Logic ER: How to Think Like an Emergency Veterinarian

A Theology of Beasts: Christians and Veterinary Medicine

Animal Anatomy Coloring and Activity Books (Canine & Feline)

School Studies

Chivalry: An Ancient Code for Our Time (high school to adults)

Chivalry: A Study for Little Knights (elementary school study)

Whatever Happened to the Vikings? (middle grade school study)

Nonfiction

Kristmenn: The Epic Saga of Viking Conversion

King: A Christian's Call to Imitate Christ's Kingship

The Tangled Eons Series

Crescent Tides

Fatal Fogs

The Stalwart Supinator: Servant of the Streets!

The Medieval Muddles

Juvament!: A Mythical Mess

The Mubblefubbles: A Toothy Tangle

Children's

Elye of Saint Gilles: A Chanson de Geste Retold (Middle Grade)

Logan the Wanderer: Adventures of a Boy out of Time (Middle Grade)

Mishaps in Miklagard: Brom Belcher's Journal #1 (Middle Grade)

Heaps of Hobs: A Fun-Filled Fable in Old English Verse (Picture Book)

Will to Start: The Extraordinary Adventures of William Moorcroft (Comic Book)

Table of Contents

WELCOME TO ANOTHER ADVENTURE IN VETERINARY MEDICINE, YOUNG VET!

You may recall your past experiences as you learned to think like a veterinarian through the cases in *Fuzzy Logic* (with a fond smile on your face, obviously). Well, get those scrubs all fluffed up in the dryer and polish off your stethoscope, because it's time to put your doctory brain back to work on a whole new set of exciting cases!

You've been a veterinarian for an entire year now. And as another colorful spring morning in the desert dawns upon blooming sage and prickly pear cactus flowers, you roll out of bed to fumble for a cup of coffee and get ready for another day of minding your fuzzy patients' many maladies. Noah's Haus Veterinary Clinic in lovely Santa Fe, New Mexico welcomes you to another busy day.

"Good morning!" the plucky, young receptionist Ms. Amy Wallender beams as you rush in through the double doors at 7:55 am. "Your first appointment is being loaded into Room 1 right now." She beams brighter. "It's Felicita Garcia with one of her strays." You beam with her at that. "But first, there's a fresh pot of coffee (and a comb) in the breakroom."

You beam brighter still, and gratefully stumble over to refill your coffee mug. As you comb your shockingly tussled mop-top, you conveniently contemplate the basic steps of inductive reasoning you learned about in the last book that taught you to THINK LIKE A VETERINARIAN!

Get your coffee (or drink of choice) and you're ready to start! **Don't forget your helpful pals**, these gray boxes... They will explain terms and concepts, give you instructions and tips to help you through the cases, and mold your young gray cells into tiny but well-tuned detective think tanks! Read 'em!

WHIRLWIND REMINDER OF THE LOGIC OF MEDICINE

Every case that comes into your clinic is a mystery! To be a good veterinarian, you need to think logically, like a detective *and* a doctor. You've been reading a book on basic philosophy and the laws of logic and have been trying to apply these laws to your work. Here's a quick overview of some important ideas you've been pondering.

Argument (sometimes called "inference"): working out a logical conclusion from a set of premises.

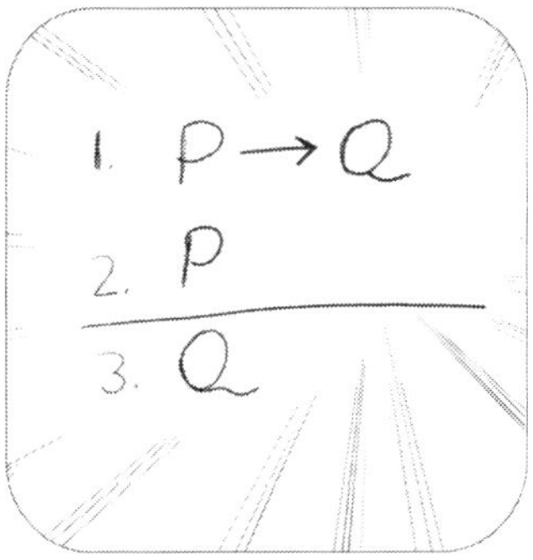

Deductive Reasoning

An argument from something general to something particular. Accepting the premise guarantees the truth of the conclusion. "Every baby is messy. This small person is messy. This small person must be a baby."

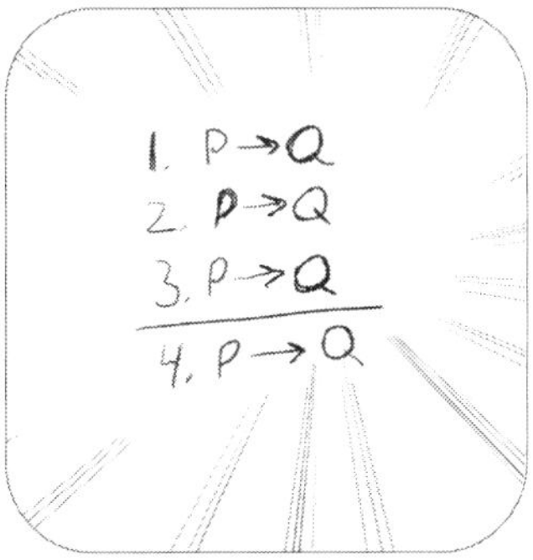

Inductive Reasoning

An argument from particular data points to make a generalization. The premise makes the conclusion more probable. "This baby's messy. Hey! This baby's messy too! Yowza! All babies must be messy."

Syllogism

An old dead Greek guy with a beard named Aristotle discovered that the roots of a valid argument lie not in its particular claims so much as in the form of the argument. Thus, in more formal logic, we tend to say an argument is valid or invalid. In general, we know an argument is valid when it's impossible for its premise to be true and its conclusion false. The proper forms valid arguments can take are called syllogisms.

The best detective-doctors use a couple of important variations of classical deductive and inductive reasoning to get to the bottom of their cases.

Refutationism: essentially, the method of coming up with a theory based on your initial clues in the history and physical exam of a case, then using tests to refute the theory (i.e., prove it wrong). Rinse and repeat within the gilded bathtubs servicing the grand palace of your mind until you discover a theory you can't refute. This is a variety of deductive logic (arguing from an initial hypothesis downward).

Bayesian Induction: besides being super-fun to say (and making you sound oh-so-clever around your friends), this mental tool is a wonderful logical labradoodle crossbreed between inductive reasoning (arguing from observations upward) and statistical probability. It acknowledges that in the real world (you know, that nasty place where disease and injuries attack your patients), there aren't really any symptoms that fit a diagnosis with 100% iron-clad accuracy. The real world gives us clues with a certain probability of occurring in a diagnosis. It uses scientific facts (based on studies), like "decreased vision occurs in 68% of visceral larval migrans (VLM) cases," to rank the likelihood that a symptom stems from a diagnosis. A doctor well-read in the latest medical studies should apply them (with at least vague probabilities like "this is more likely") in his mind while evaluating possibilities in cases. That's how we reason using "evidence-based medicine"—making our decisions based on published statistical evidence (instead of hearsay or experience).

"In medicine, we make a list of the possibilities, and cross stuff off with tests until we get an answer."

In the simplest terms, a smart veterinarian-detective approaches a case using the clues in the history and physical exam to make a list of possible causes of the symptoms. This is called a differential list. Then you rank that list of possible diagnoses from the most likely to the least likely in your brain. Then you use tests (like bloodwork and X-rays) to cross things off the list that are least likely until you come to the most probable diagnosis. Thus, in medicine, we make a list of the possibilities and cross stuff off with tests until we get an answer.

And Bob's-yer-uncle, you've solved another case!

"'Contrariwise,' continued Tweedledee, 'if it was so, it might be; and if it were so, it would be; but as it isn't, it ain't. That's logic."

—Lewis Carroll

Differential list: the list of possible causes ("differentials") for a patient's problems. List of diseases that might make the patient sick.

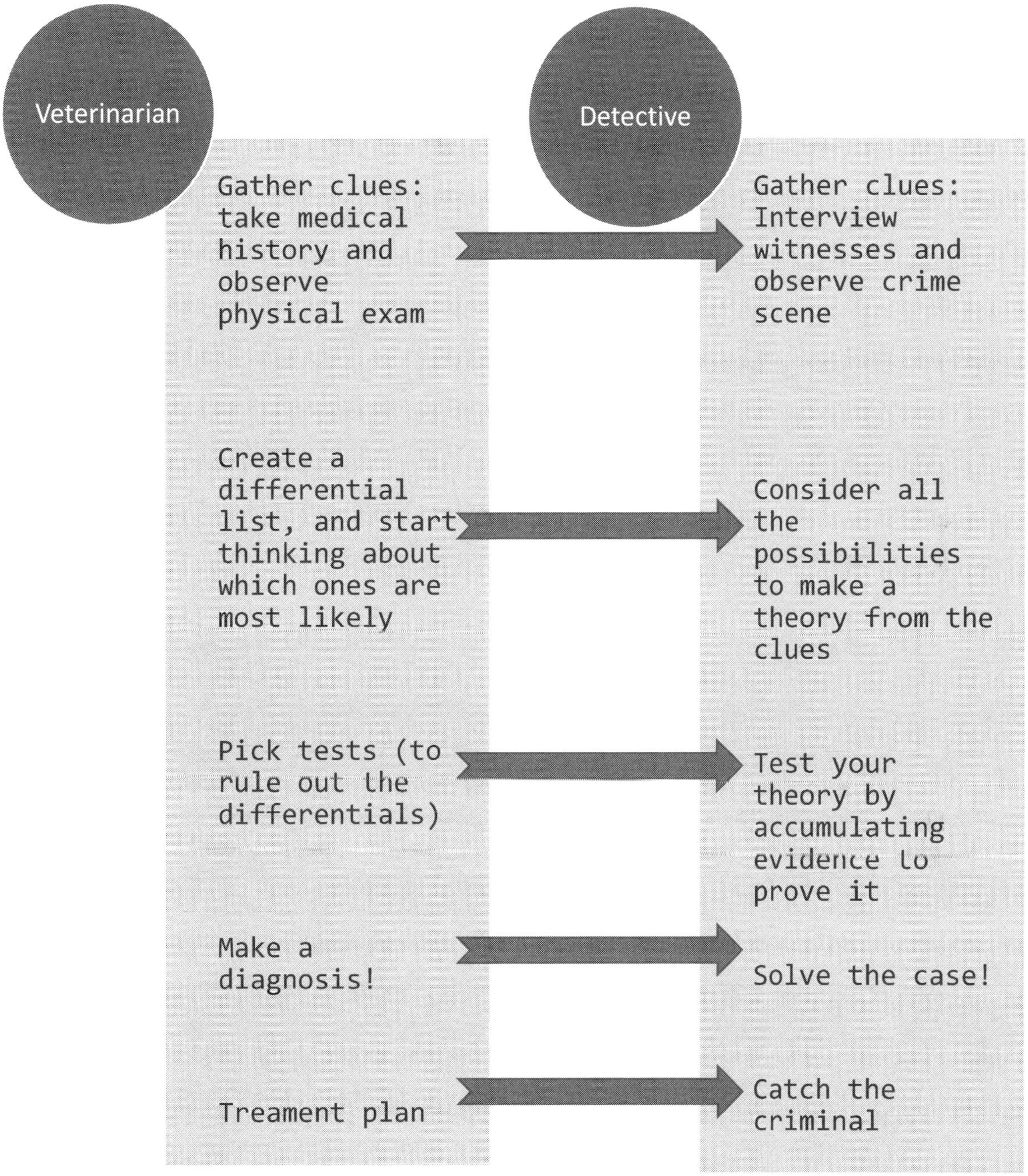

You may remember this diagram from Fuzzy Logic 1.
It remains super-sweet, and you should peruse it again!

Case #1

> If you get stumped by a big, scary medical term, do not fret! You can peruse the definitions of terms at the bottom of the pages when they first appear. You can also consult the handy, comprehensive glossary at the back of the book.

A scruffy-looking terrier-ish thing blinks at you from the arms of Felicita Garcia. You smile at the matronly Mrs. Garcia as you enter the room and don a pair of exam gloves.

"It's good to see you again, Mrs. Garcia." You aren't merely being polite. This warm-hearted Hispanic grandmother is a favorite client at the clinic. She's basically become your adopted grandma. And by this time, you've learned to appreciate your friendly clients immensely, as a wonderful antidote to the less-than-friendly ones. Today she has a little boy with her you've never met. "We all loved those cookies you brought last week. Thanks again! And who's this?"

"This is my cousin's grandson, Samuel. He was bored at the restaurant and needed an outing."

Samuel's dark eyes gleam, and he grins broadly when you give him a fist bump. He's about 5 years old and seems bursting with curiosity about everything in the office. "*Mucho gusto*, Samuel" you say.

"Yeah. You too," he replied. The pup licks his toes, and he giggles a bit. You notice his speech is just a little slurred and comes slowly. He goes back to playing a game on his phone, and you go back to talking with Abuela Garcia.

"You've found another stray, I heard. Does this fella have a name yet?"

"I was thinking of Gustav Mauler because I found him outside a symphony concert the other night." You and Samuel grin at the ridiculous name before you ask what's ailing Gustav today. "He's had diarrhea for the past few days, and he vomited once."

"Blood in the diarrhea?"

"No."

"Still eating?"

"Ay, *Chico*, he eats all right! Seems like he's always hungry. But he still seems skinny to me. Does he look skinny to you?"

"Maybe so…" you admit.

Gustav Mauler blinks at you again and licks his lips as you start your exam. He does have a hungry sort of look about him.

"You know skinny doesn't ever sit well with me. Must be what comes of working at Hector's restaurant, and constantly filling people full of savory green chiles over smothered burritos with frijoles." Felicita works at La Chuleta, a local restaurant owned by her cousin Hector Melendez.

You feel hungry now and make small talk to keep your mind off those delicious chiles. "How are things at the restaurant?"

"*Bien, bien.* Slow lately. I really don't know how Hector keeps the place going."

"Yeah?"

"But he always pays his employees well and is a good boss. He pays me even better, because…" her chatter unexpectedly stops. "But never mind. What do you think about little Gustav?"

Fortunately, your multitasking little gray cells have already analyzed the problem, and your keen sense of observation has already noticed several important things…

Step 1 – Organize the History

> In this book you'll learn to break down the logic of each case into a few simple steps. **Step 1** is to organize the information (history) so you don't miss anything important. This consists of A) the *signalment*: a fancy word for a description of the patient and B) *presenting complaints*: the problems the patient comes in for.

Signalment:

Terrier mix

Male intact

Estimated 2-3 years old

Presenting Complaints:

Vomiting

Diarrhea

Thin—not gaining weight

Signalment: the descriptive characteristics of the patient.

Step 2 – Gather Clues

> Gathering clues will be **Step 2** for each case. Remember that a veterinarian is like a detective and needs to gather clues early in a case to figure out what disease is making the patient sick. We do this by a *physical exam* (PE) first, then by running tests based on clues from the history and PE.

Physical Examination (PE):

Abdominal palpation: Abdominal palpation is WNL (Within Normal Limits). This means it's pliable, no splinting, no obvious masses noted, no distension noted, etc.

Attitude: BAR (bright, alert, responsive).

Cardiac (heart): WNL. No murmur or arrhythmia noted.

Eyes: WNL.

Ears: Waxy debris and dirt, otherwise canals appear clean.

Integument (skin): Slightly decreased skin turgor (estimated 5% dehydration).

Lymph Nodes: WNL.

Musculoskeletal: WNL.

Neurologic: WNL.

Oronasal (mouth and nose): WNL.

Pulmonary (lungs): Lungs sound clear. No increased respiratory effort noted.

Urogenital (urinary and genital organs): WNL.

Vitals: Mucous membranes are pink and tacky, Capillary refill time (CRT) <2 seconds.

BAR: Bright, alert, and responsive.
Capillary refill time (CRT): capillary refill time. A simple way to check blood flow (perfusion) by pushing down on the gums and counting how long it takes to pink up again in that spot. It should be less than 2 seconds in a healthy beast.
Distention: gastric distention means the stomach is dilated and bloated.
Integument: a fancy word for skin.
Lymph node: important structures for the immune system. "Hubs" of white blood cells and lymphatic drainage, to fight infection in specific places of the body. Small and bean-shaped, like the author of this book.
Mucous membranes: the wet bits of a critter at the junction of skin and mucosa (like the gums in the mouth).
Oronasal: the mouth and nose.
PE: Physical Examination.
Skin turgor: the tension of the skin, how well it falls back down when you tent it upward. This is a way we check the hydration status of a patient.
Splinting: tensing of the abdomen because of pain or fear.
Urogenital: urinary and genital organs.
WNL: Within normal limits.

Differential Diagnoses (the possible diseases):

- Congenital disease (some structural problem he was born with)
- Gastroenteritis
- GI foreign body/obstruction
- GI infection
- GI parasitism
- Hepatic dysfunction
- Intussusception
- Neoplasia
- Pancreatitis
- Toxicity

Differential Diagnoses: For every case you must have a running list of possible solutions (diagnoses) that you are working through, to reach a single, most likely diagnosis by process of elimination. In other words, it's a list of every possible cause you can think of for the problems. It essentially makes every case a really big multiple-choice question. This is the basic thought process in medicine. You use clues like tests and observations from PE and history to rule out the possibilities (*differentials*) until there's only one logical one left. And that is your diagnosis.

Blood gas analysis: a blood panel specifically designed to check chemistry related to respiration and blood pH (like oxygen and bicarbonate levels).
Clotting times: a test of the ability of blood to clot properly.
Congenital: describing a condition a patient is born with.
Differentials: the list of possible causes for the patient's problems. List of diseases that might make the patient sick.
Fecal direct saline exam: looking at a smear of poop in solution on a microscope (usually to examine the bacteria).
Fecal floatation: a test for parasites run by floating a gram of poop in a special solution for 10 minutes, then looking for parasite eggs under a microscope.
Foreign body: a thing that's not where it should be, like something indigestible inside a loop of bowel.
Gastroenteritis: inflammation in the GI tract (often caused by bacteria or viruses).
GI: gastrointestinal. The stomach and intestines (supported by the spleen, liver, and pancreas).
Hepatic: related to the liver.
Intussusception: an intestinal obstruction caused by a piece of intestine telescoping in on itself and getting stuck into a clump.
Neoplasia: cancer.
Pancreatitis: inflammation in the pancreas, which is an organ that makes hormones and helps digestion.
Thoracic: related to the chest (also called the thorax).

Decision time!

> Here's where the rubber meets the road! Veterinarians have to make many choices for every case, each of which will have its own set of side effects and consequences (to the patient, the client, and the vet). If you're the sort of person who can never make a decision, then medicine may not be the career for you!
>
> In this book you will choose two things for each case:
> 1) What tests to run.
> 2) What treatment to start.
>
> Your choices should be based on what you read in the history and Physical Examination (PE), and what diagnoses you deduce from them. The answers (and more importantly, the thought process algorithm) will be presented as you read further in the book.

What tests do you want to run first?

Here's what you can do in your clinic:

- ✓ Basic bloodwork (a chemistry panel and CBC)
- ✓ Blood gas analysis
- ✓ Clotting times
- ✓ Cytology
- ✓ Fecal floatation and direct saline exam
- ✓ FeLV/FIV test
- ✓ Giardia test
- ✓ Heartworm test (with rickettsial diseases)
- ✓ Leptospirosis test
- ✓ Pancreatic lipase
- ✓ Parvo test
- ✓ Pregnancy test
- ✓ Radiographs (X-rays)
- ✓ Skin scraping
- ✓ Thyroid hormone level (T4)
- ✓ Ultrasound
- ✓ Urinalysis
- ✓ Wood's test or DMT plate

> Heartworms are real worms that live in the right side of an animal's heart and cause heart failure!

Chemistry panel: a blood panel that usually has basic information on the function of the liver, the kidneys, electrolyte levels, protein levels, and blood glucose.
Cytology: looking at cells under a microscope. If you pick this, you have to pick where to get the cells from. Will it be scrapings from the skin? Or cells from a tumor with a needle (fine needle aspirate)? Or cells from a swab of ear gunk or snot?
DMT plate: a special petri dish you can grow dermatophytes (ringworm) on.
FeLV/FIV: feline leukemia virus and feline immunodeficiency virus. These contagious diseases suppress the immune system in cats.
Giardia: a microscopic parasite that looks like a flying saucer with googly eyes. It causes a nasty infection in the intestines that produces diarrhea and vomiting.
Leptospirosis: a bacterial disease that causes kidney and liver failure, often contracted through contaminated water (much like giardia). This bug looks like a coiled, swimming pigtail.
Pancreatic Lipase: a test for pancreatitis. There's an fPL (feline pancreatic lipase) and a cPL (canine pancreatic lipase).
Parvo test: a test for canine parvovirus, which is a virus that attacks the GI tract in puppies.
Radiographs: X-rays. A way to use radiation to take pictures of an animal's insides.
Rickettsial diseases: some infectious diseases carried by ticks (lyme is one of them).
Ultrasound: a way to use sound in ultrasonic frequencies to view pictures of an animal's insides.
Urinalysis: a test on urine, generally looking for infection or crystals from bladder stones.
Wood's test: shining a blacklight on the skin to look for ringworm (a fungal disease of skin). Most cases of ringworm glow.

What tests do you want to run?

What tests did you pick? I picked: some **basic bloodwork**, **fecal testing** (floatation and direct saline exam), and a **giardia test**.

Give yourself a gold star if you picked those out too! Other tests that might make sense too, like radiographs, pancreatic lipase test, or even rickettsial testing. If you picked them, give yourself a rusty copper star, which is only a small downgrade from a gold one. Don't feel bad if you picked tests I didn't! Testing for something you didn't need to is not such a bad thing. You miss things more often by not looking for all the clues than you do by looking and getting a boring result.

Your amazingly strong-stomached technician Fred Kawalkovitch takes the patient outside to collect a stool sample for testing. Before going outside to try talking Gustav Mauler into pooping on your nice, clean parking lot, Fred pulled some blood from the patient with the help of your kennel tech Katalina Krumm (stop chuckling - she can't help her name, it was her parents' fault). Your boss recently bought a spiffy new in-house lab analyzer, and it accepts the proffered blood into its cybernetic mouth, and soon spits out the results of a general blood panel. Turn the page to behold and wonder at it!

Client: Garcia, Felicita	Species: Canine	Sex: M
Patient: Gustav Mauler	Breed: Terrier c	
Patient ID: 026567		

When you look at bloodwork, run your eyes over this "flag" column first: this clues you into which results were high (above the reference range) and which were low (below the reference range).

There is a more detailed explanation of the bloodwork panels in Appendix A.

CBC

TEST	RESULT	FLAG	REFERENCE RANGE[i]	
WBC	17		6.0-17.0	10x3/µL
Neutrophils	8		3-11.4	10x3/µL
Monocytes	0.1		0.15-1.35	10x3/µL
Lymphocytes	2		1-4.8	10x3/µL
Eosinophils	1.89	H	0.11-0.75	10x3/µL
Hematocrit	57	H	37-55	%
MCV	67		60-77	fL
Platelet Count	302		200-900	10x3/mm^3

White blood cells. These go up with inflammation, to fight infection.

These are specific types of WBC.

Eosinophils are a type of white blood cell that tend to increase with parasites or allergies.

He's a bit hemoconcentrated (high hematocrit = more red blood cells per volume). Dehydration often causes this sign.

Client: Garcia, Felicita	Species: Canine	Sex: M
Patient: Gustav Mauler	Breed: Terrier cross	
Patient ID: 026567		

Chemistry Panel

TEST	RESULT	FLAG	REFERENCE RANGE[ii]	
ALT	50		8.2-57	IU/L
ALP	25		10.6-101	IU/L
GGT	5		1.0-9.7	IU/L
tBili	0.3		0.1-0.6	mg/dL
Glucose	84		62-108	mg/dL
Cholesterol	113		116-254	mg/dL
Total Protein	7.7	H	5.5-7.5	g/dL
Albumin	2	L	2.6-4.0	g/dL
Globulin	5.5	H	2.1-3.7	g/dL
Calcium	8.2		8.7-11.8	mg/dL
BUN	24		8.8-26	mg/dL
Creatinine	1.1		0.5-1.6	mg/dL
Phosphorus	5		2.9-6.2	mg/dL
Sodium	159	H	140-154	mEq/L
Potassium	3.9	L	3.8-5.6	mEq/L
Chloride	116	H	102-117	mEq/L

Liver (ALT, ALP, GGT, tBili)

Albumin: a type of protein made in the liver and excreted in stool and urine. Low albumin can happen from liver disease (not making enough) or kidney and GI disease (excreting albumin too fast to keep up).

Globulin is a type of protein produced by the immune system to fight infection.

Kidneys (BUN, Creatinine, Phosphorus)

Electrolytes (Sodium, Potassium, Chloride)

There are mild electrolyte imbalances in this patient. This is likely from vomiting and diarrhea, which excretes potassium. Dehydration makes a critter's salt content (i.e. Na and Cl) elevated.

OK. Time to take those numbers, scan them with your brain's eagle eye from on high, and try to look at the entire picture to understand them… Look for patterns left by disease processes.

We have:

- Elevated hematocrit (hemoconcentration) and sodium and potassium. Those are clues our patient is dehydrated.
- Low albumin (we're spilling it out in the diarrhea).
- Elevated eosinophils (white blood cell count) right at the high end of the range, and elevated globulins. These are clues of infection, specifically a parasite infestation.

Usually we use the term "infestation" instead of "infection" for being riddled with parasites…

That's the more practical way to look at this bloodwork.

The more technically correct way (and a method to avoid missing things) is to make a differential list for each of the abnormalities on the bloodwork. That's because, out here in the real world of complexity, there are all kinds of causes for everything you find out of place.

Kind of like this…

Hemoconcentration

- AHDS
- Dehydration
- Laboratory error (like leaving the blood sitting out too long before running the sample)
- Polycythemia vera
- Renal cancer

High Sodium

- Decreased water intake
- Dehydration
- Diabetes
- Dietary (eating increased salt amounts)

AHDS: Acute Hemorrhagic Diarrhea Syndrome (sometimes called HGE, Hemorrhagic Gastroenteritis). This is a condition in dogs that makes them poop blood, while the blood is concentrated in their veins.
Albumin: a type of protein made in the liver and excreted in stool and urine.
Eosiniphils: a type of white blood cell that tends to increase with parasites or allergies.
Globulin: a type of protein produced by the immune system to fight infection.
Hemoconcentration: elevated hematocrit (high red blood cells).
Polycythemia vera: a very rare disease where the blood is just super concentrated. This is one of the very rare cases we can still use medical leeches for! There are leech farms in the world for things like this…
Renal: related to kidneys.

- Diuretic medication (help rid the body of sodium and water)
- Hormonal (diseases affecting pituitary glands or hypothalamus)
- Kidney failure
- Pancreatitis

High eosinophils

- Allergies
- Infiltrative disease
- Parasites

Hypokalemia

- Diabetes
- Diarrhea
- Dietary
- Digestive disorders
- Diuretic medication
- Hyperthyroidism
- Kidney failure
- Vomiting

Remember!

You get down to the right answer by using clues to eliminate the improbable! That's why you'll find a lot of detective quotes in this book too.

And on it goes… I'll stop listing differentials for each bloodwork abnormality because that might take up most of this book. Indeed, there are books that consist only of differential lists for medical problems (and they're very useful).

It's also true of tests other than bloodwork! Abnormal findings on radiographs and ultrasounds (and all the others) each have a differential list behind their results—there are lots of possible causes for each abnormal finding. I'll steer you toward the most likely causes, but just remember as you think through a case to consider as many differentials (possible causes) of every find you come across. That's how clinical practice usually works: I'm coming up with short differential lists in my head whenever I look at diagnostics and trying to decide which possible cause (differential) is likely enough to take seriously.

Hyperthyroidism: a disease process in which the thyroid gland produces too many hormones, affecting the body's systems.
Hypokalemia: low potassium levels.
Hypothalamus: gland in the back of the brain that controls hormones.
Infiltrative disease: a condition where cells invade places they shouldn't. Often we talk about this in the intestines in particular, and often with types of cancer.

Well, it looks like Fred has finished readying the stool sample that Gustav Mauler so kindly provided! He's made a fecal direct smear by slipping a bit of that useful poop onto a microscope slide, then adding a drop of saline and a coverslip. Let's take a look into the microscope, shall we? You dial from 10x down to 40x and all the way to 100x, getting things smaller and smaller, until you can get a good feel for the bacterial population of the stool.

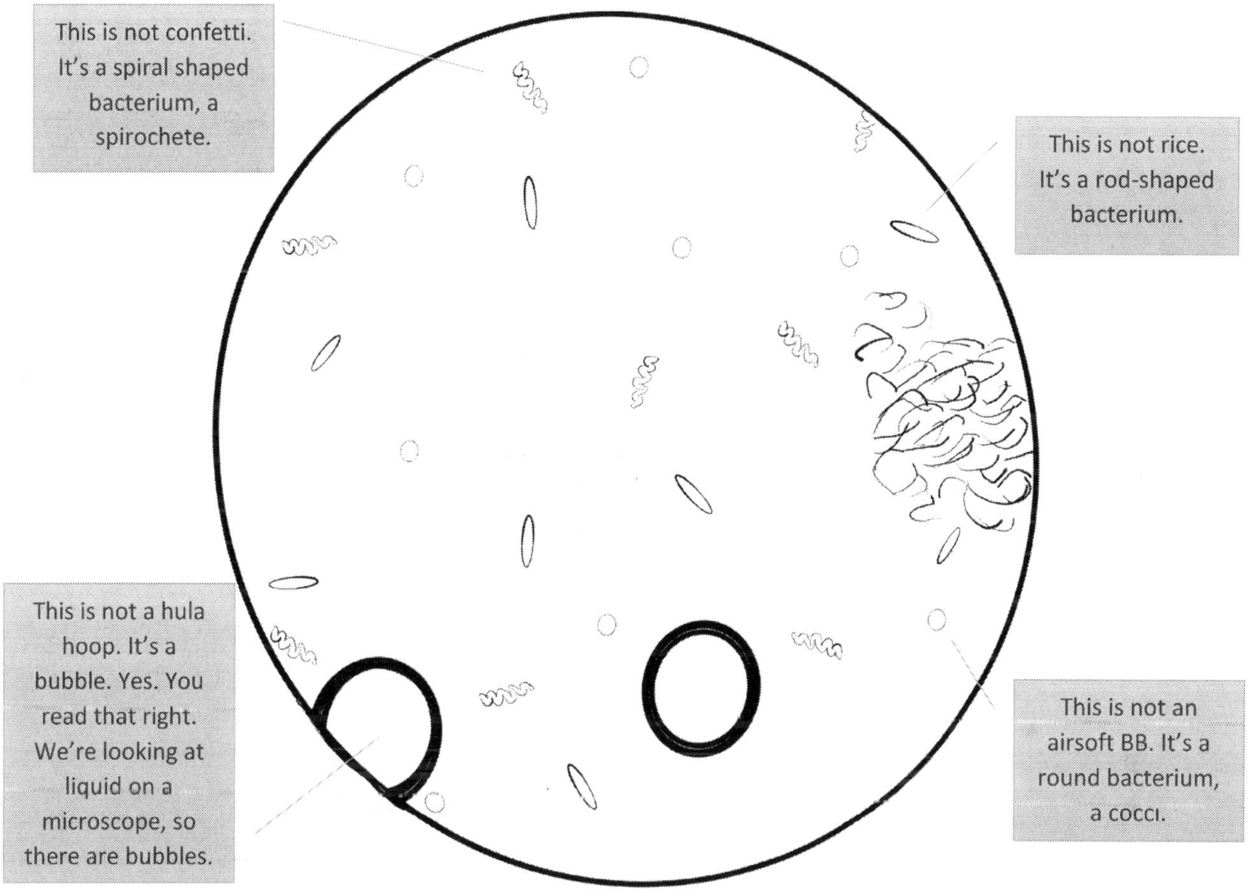

It may come as no surprise that normal poop is full of bacteria. There are conditions that'll mess up the normal bacteria in the intestinal tract and fill it with bad bacteria. Characterizing the bacterial population is a major point of running a fecal direct smear. These little beasties come in three basic shapes: spheres (called cocci), rods (called bacillus), and spirals (in stool the spirals are usually called spirochetes).

I'd say this stool has more round guys than rod guys (normally there would be about equal proportions of the two). It looks like a heavy bacterial population, though, with lots of spirochetes zipping around… On paper you can't see them zip, but in real life they look like a drunk track team, dashing and spiraling all over the place. And on a stool sample, they usually show up when things are out of balance.

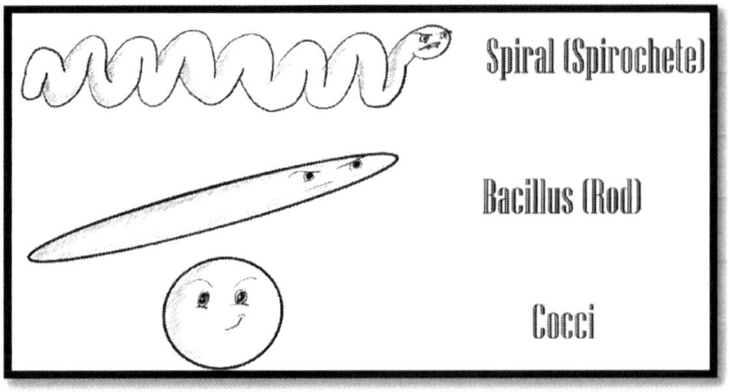

Fred has floated around 1 gram of stool in a special solution against a coverslip for 10 minutes and now puts the coverslip onto another slide. This solution is designed to make any microscopic parasite eggs float up to the coverslip so we can see 'em when we stick it to a slide.

Let's take a gander through the microscope yet again!

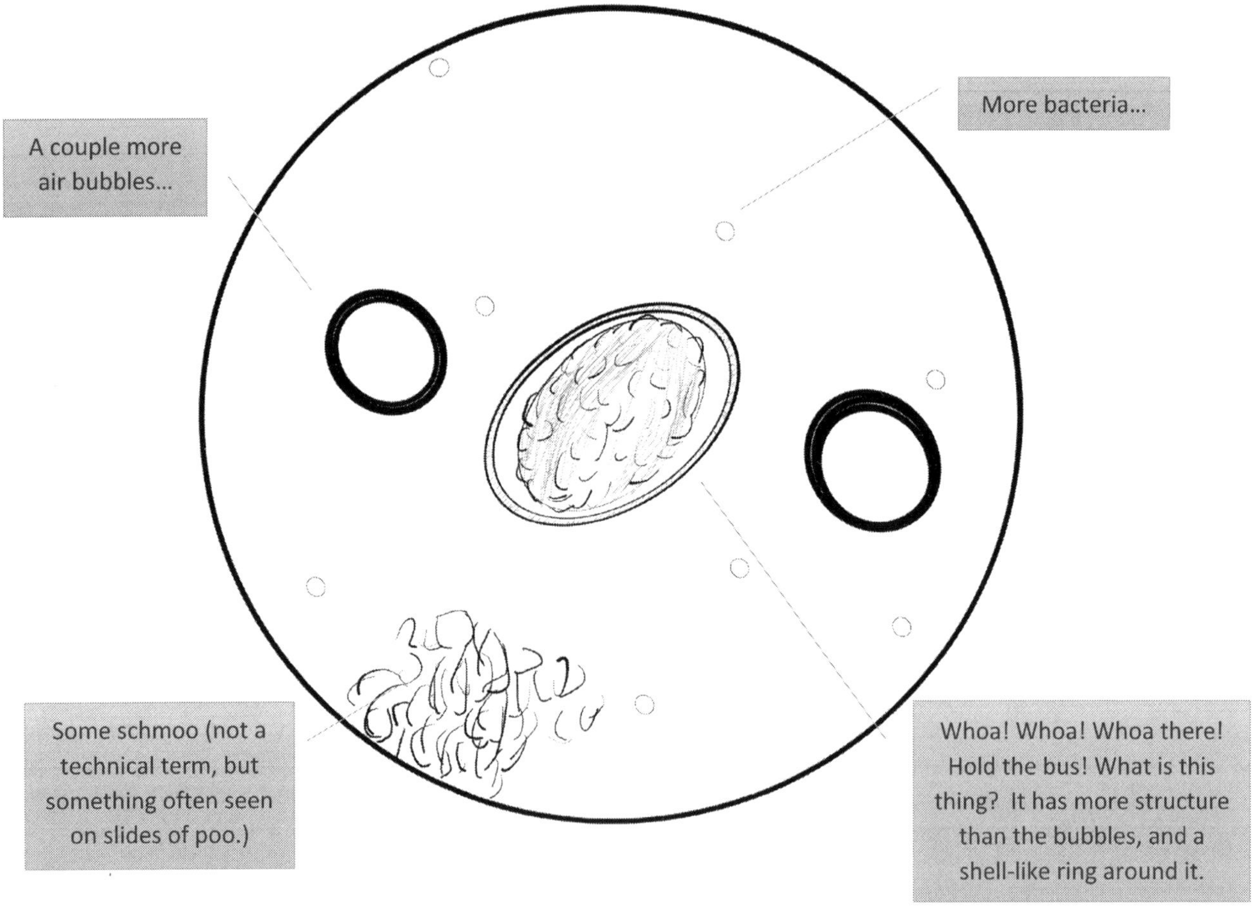

So, about that shell-like ring around the eggy-looking guy you read about in the gray box (you are reading the boxes, right?)… That shell-like thing is, well…a shell. That is a parasite egg. More specifically, it is an ovum of *Toxocara canis*, a species of roundworm (also called ascarids) that infests dog guts sometimes.

You do a dance of glee. You have answers!

And your diagnosis is?

____

STOP!!!!!!

Before you turn the page,
try to ID the culprit…
What's the most likely cause of this patient's signs?
Go back a few pages to the disease differential list.

I see giardia, leptospirosis, and some other bacteria called clostridia in this random illustration. That is one angry, angry GI tract with some serious diarrhea! These are things you might see on a fecal direct saline exam.

If you answered **GI parasites**, you are totally correct! If you answered **roundworm infestation**, you're even more correct!

But wait, there's more…

What about that bacterial population being out of proportion, and crowded with zoomy little spirochetes? Those are signs of imbalanced intestinal bacteria, consistent with gastroenteritis. If you answered **GI parasites and gastroenteritis**, you're right as rain! And rain is always right.

"Ascarids" is the fancy name for roundworms. These are a broad group of intestinal parasites that attack all kinds of animals. Specific species of ascarids mooch off a specific species of animal intestines, but fortunately, their egg structure is all variations on a theme. Dogs and cats can get roundworms by being born with the bugs or by accidentally ingesting their microscopic eggs shed in another animal's poop,[1] or by eating other critters already infested with roundworms.[2]

Roundworms are, as their name implies, round. They look a lot like wiggly, living spaghetti noodles. And they live and thrive and grow in the intestinal tract, mooching off their hosts and using up the hard-earned energy their hosts get from eating. Their wiggling and munching inside of a dog causes diarrhea and sometimes vomiting. Gustav Mauler's bloodwork reflected his body's response to these wriggly hitchhikers, in the form of eosinophils (white blood cells to fight them) and globulins (antibodies to identify them). The rest of the bloodwork abnormalities were reflections of the vomiting and diarrhea the parasites caused.

Gustav's tests showed evidence of gastroenteritis secondary to the worms. They've caused a shift in the normal bacterial population of poor Gustav Mauler's once healthy intestinal tract, which let bad and opportunistic bacteria thrive and overgrow. And that produces more response from the body causing inflammation (and more vomit and diarrhea).

You make Samuel's day by showing him the worm eggs under the microscope. Why do little boys universally get excited about poop?

He asks an insightful question. "What are we going to do to fix Gustav?"

You've got an intimidating arsenal of medical weaponry at your disposal here at Noah's Haus Veterinary Clinic. But you'll have to pick the right one for the job. How do you make this pup healthy again?

Ascarids: the fancy name for roundworms and a broad group of intestinal parasites that attack all kinds of animals.

[1] Let that be a lesson to you, gentle reader… DON'T EAT POOP!!!!!!!!!!!!!

[2] Also, don't eat mice. If you really feel you have to eat a mouse, cook him very well.

Choose some from this list of treatment options.

a) Abdominal exploration
b) Antibiotics
c) Deworming medication
d) Fluids
e) Nausea drugs
f) Prescription food
g) Probiotics
h) Steroids
i) A vigorous shaman rain dance

Surgery to look for problems in the abdomen.

Drugs that slaughter bacteria en masse.

Supplements of good bacteria.

Synthetic steroids (the kind not made by your body) are usually given in doses to either fight inflammation or suppress an overactive immune system.

What treatment plan are you going to go with?

"All opinions are not equal. Some are a very great deal more robust, sophisticated and well supported in logic and argument than others."

—Douglas Adams, *The Salmon of Doubt*[iii]

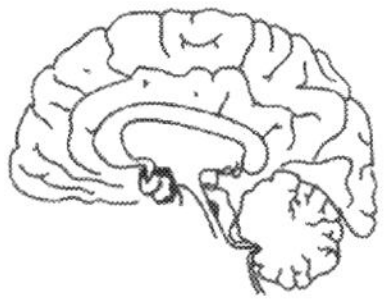

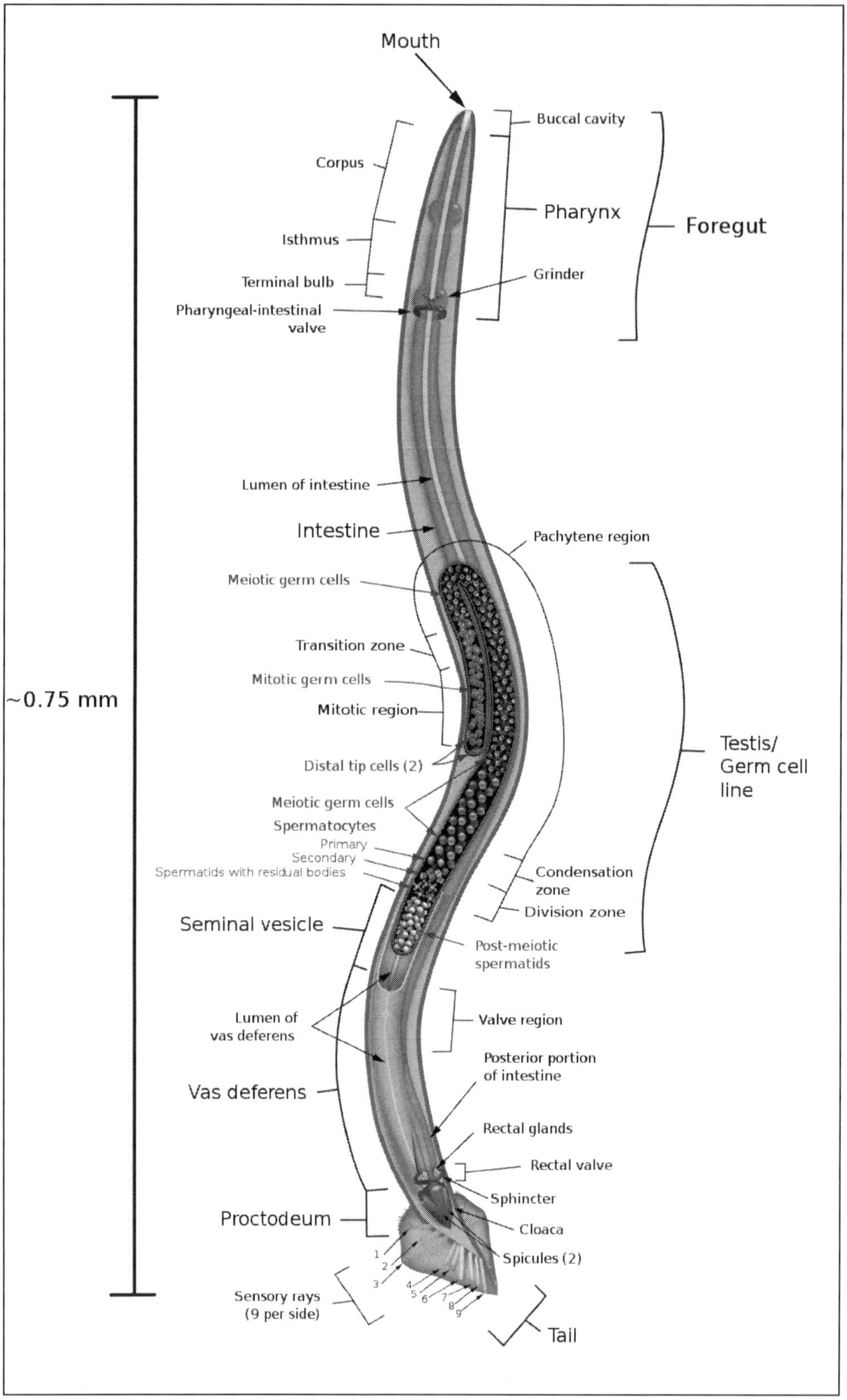

This is an anatomy diagram of a non-parasitic roundworm (a friendly cousin of the ascarids, called a nematode).

You recommend treatments **b, c, d, e, and g.**

Nice job! You've appropriately decided to get rid of the worms with a **dewormer**. That should do it.

Specifically, you pick pyrantel pamoate, and do some cool math to get a dose. You wisely warn Mrs. Garcia that she'll probably see dead worms in Gustav Mauler's stool for a few days as the worms die and lose their handy foothold in his doggie guts.

Another good idea is to use a GI **antibiotic** (metronidazole is the most common one) and a **probiotic** spaced out from it to help reset the intestinal population of bacteria and treat the gastroenteritis. "Stop that crazy disco party in Gustav's guts, you wild bacteria! Get out of here, so the nice quiet ones can get back to work!" You decide to give **fluids** subcutaneously (injected under the skin in a bubble near the shoulders), which Gustav Mauler can absorb at home. And because he's vomiting, you give your grateful patient an injection of **nausea medications**.

"That should do it!" you proclaim proudly, as you see Mrs. Garcia and her little friend to the door. "Just let us know if your fella isn't right as rain in a couple of days."

"Bueno! Rain is always right." She seems happy, and you feel confident more delicious cookies could be in your future.

"Bye! Thanks, mister doc!" Samuel is happy too and hands you a picture he drew of Gustav Mauler. He's a pretty good artist! You've successfully entertained him. A good start to the day!

Your next three appointments are vaccines only and are accomplished without excitement. A first puppy exam is next, during which you get to cuddle with a minuscule poof-ball that will someday grow into a Pomeranian (i.e., an adult minuscule poof-ball). Your tongue can't help speaking baby talk to him, even though your brain tells it to stop. That's a reflex you sometimes pick up in vet school. Just when you start to think this day may become monotonously pleasant, Fred grins at you and hands you another chart.

"Got something more interesting for you next, Doc…"

Probiotic: a supplement of good bacteria that helps establish normal flora in the gut.

> **Formal logic** uses symbolism, like math for philosophers. Here's a page explaining the meaning of some of the common symbols used by logicians. You'll see these periodically in the following pages, so earmark this page to come back to if you need it…

$\rightarrow$	"If… then"
$\neg$	"Not"
$\vee$	"Or"
X	"All x are ______" (Used in universally quantified statements)
$\exists$	"There is as least one ______ that…" (Used in existentially quantified statements)
$\Box$	"Necessarily" (in modal logic)
$Pr(A/B)$	"Probability of A given B…" (for inductive logic)

> *Existentially quantified statement:* a way to say that at least one of ______ exists out there.
> *Modal logic:* a branch of logic distinguishing modes of truth (like necessary vs. conditional truth).
> *Universally quantified statement:* an "if… then" statement phrased like "all ascarids are worms."

Case #2

Fred hands you a sticky note with his messy writing scrawled thereon. "It's a 13-year-old corgi with GI upset. Here's a list of the meds he's on right now. I can get the doses when I get a more detailed history." You tell him you'll get the history for this one since three more clients walk in, and Fred is needed to help triage and load them into rooms.

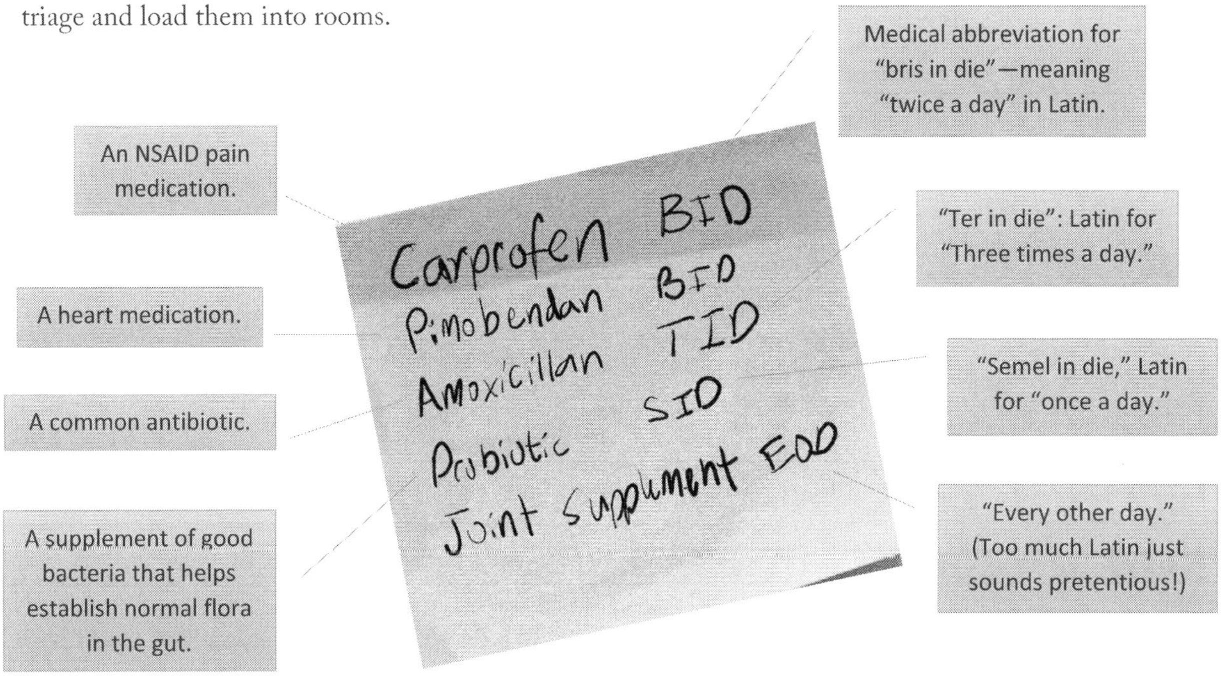

A tricolored corgi named Carl meets you in the exam room. He doesn't appear to be feeling well at all. Fortunately, Carl's brought along a friend (his owner, Bernard Gallagher II) to translate for him and pay his doctor bills. You tell Bernard he looks familiar as you shake his hand.

"Yeah. I've seen you several times at the bookstore I work at. You always buy comic books." His tone is just a tad accusative and you suspect he's thinking something like, *Just my luck to get the kid doctor who sits around all day reading comic books.*

You wisely move the conversation along to ask him about Carl's problems.

Antibiotic: a medication that kills or inhibits the growth of bacteria.
BID: twice a day.
EOD: every other day.
NSAID: non-steroidal anti-inflammatory medication. This is a type of pain medication that helps with inflammation, which is not a steroid. Aspirin is a common human one.
SID: once a day.
Triage: a quick assessment to decide how critically sick an animal is, for ordering which patients need to be seen first.

Ooh look! This pup is guarding a logical syllogism. These "laws of logic" will pop up here and there. Apply them to cases!

Modus Ponens

1. $P \rightarrow Q$
2. P

---

3. Q

1. "If antibiotics are given to cat bite abscess cases (P), there's a 95% chance they'll heal (Q)."
2. "Antibiotics we're given to this cat bite abscess case."
3. "There's a 95% chance he'll heal!"

"He started acting sick three days ago. Just less energy, you know. Vomited a couple of times two days ago. I was busy reading Tolstoy and Dostoevsky and didn't have time to bring him in. But he was still vomiting yesterday, and started having bloody diarrhea too."

"Still eating and drinking?"

Bernard gives you a funny look. "Every day. I went to this Swedish Meatball restaurant downtown for dinner yesterday. And I drink occasionally. Mostly white wine, but sometimes I'll settle for a nice rosé. The other day I bought this exquisite 1985 Henri Jayer Cros Parantoux from a guy I met at…"

Carl whimpers and you explain you meant to ask if *Carl* is still eating and drinking.

"Ah… Thought that was an odd question," Bernard seemed unruffled. "Erm…and maybe forget I mentioned the Parantoux. Get my drift?" He winks at you, and you have no earthly idea what he's talking about, so you repeat your question. He thinks for a moment. "No. Carl's not eaten in two days, by my estimation. And he stopped drinking yesterday."

"Any toxin exposure?" You realize you should have rephrased that the minute it leaves your mouth.

"Not much. I handle some rat bait at the house once a month and sometimes spray my weeds. Some toxic personalities come into the bookstore pretty regularly, ha ha. And if you count alcohol… Oh, you mean Carl again, don't you? Not that I know of."

"Has *Carl* eaten anything weird lately?"

Bernard stops to consider what might be weird for a dog to eat. "Well, he ate a dead rat a week ago. And he got into the trash mebbe four days ago. Nothing in there but some remnants of a Caesar salad and chicken cordon bleu."

Aha! the facetious part of your mind whispers to you. *Fancy foods that literary people eat... Obviously goes well with that nice rosé, and maybe duck confit...*

But you make that ornery mind turn back to the task at hand. You ask about the amoxicillin Carl's on, and Bernard is helpful. "It was scripted here about two years ago. I was worried he'd get an infection after he got into the trash and started it right afterward. Look, I've got to get to a cheese tasting tonight, and then go to book club. If you can just tell me what's wrong and get Carl some pills, that'd be great..."

You move on to your physical exam, simultaneously organizing the clues in your brain as your deft fingers move in that masterful, multi-tasking way you are so very brilliant at.

Step 1 – Organize the History

Signalment:

Male neutered

Welsh Corgi

13 years old

Presenting Complaints:

Inappetence

Bloody diarrhea

Lethargy

Vomiting

Step 2 – Gather Clues

Physical Examination:

Abdominal palpation: Tense and painful on abdominal palpation, especially the cranial abdomen (the front part, by the ribs). No masses or foreign body noted. No distension appreciated.

Attitude: Quiet, depressed, lethargic, and weak.

Cranial: positioned on the body closer to the head than the tail.
Inappetence: refusing to eat.

Cardiac: Murmur noted (grade 2/6 on the left side).

Ears: WNL. Canals appear clean bilaterally.

Remember, WNL means "Within Normal Limits."

Eyes: WNL.

Integument: Severely decreased skin turgor (estimated 10% dehydration). No bruises noted on skin.

Lymph Nodes: WNL.

Musculoskeletal: WNL.

Neurologic: WNL.

Oronasal: WNL. Mild dental calculus.

Pulmonary: Lungs sound clear. No increased respiratory effort noted.

Urogenital: WNL.

Vitals: Mucous membranes are pink and tacky, CRT around 2 seconds, no petechiation noted on gums.

Based on the information you have so far, you come up with this list of possible causes for Carl's symptoms:

Differential Diagnoses:

Gastroenteritis

Pancreatitis — An inflamed, angry pancreas (organ that makes digestive juices and enzymes)

GI foreign body/obstruction — Something stuck in the intestines or stomach

GI parasitism

Heart disease (valve diseases, heart muscle problems, etc.)

GI infection

Toxicity — Rat poison might be of particular concern here, given the history...

Medication side effects

Congenital disease — An organ problem he was born with

Hepatic dysfunction — Failing liver

Renal dysfunction — Failing kidneys

Neoplasia — Cancer

Intussusception

Heart Murmur: a swishy sound on the stethoscope when listening to the heart. This is usually from a process that swishes the blood around (creates turbulent flow), like a problem with a heart valve.

Petechiation: teeny little bruises on the skin or gums.

Decision time!

What tests do you want to run first?

Here's what you can do in your clinic:

- ✓ Basic bloodwork (a chemistry panel and CBC)
- ✓ Blood gas analysis
- ✓ Clotting times
- ✓ Cytology
- ✓ Fecal floatation and direct saline exam
- ✓ FeLV/FIV test
- ✓ Giardia test
- ✓ Heartworm test (with rickettsial diseases)
- ✓ Leptospirosis test
- ✓ Pancreatic lipase (fPL or cPL)
- ✓ Parvo test
- ✓ Pregnancy test
- ✓ Radiographs (X-rays)
- ✓ Skin scraping
- ✓ Thyroid hormone level (T4)
- ✓ Ultrasound
- ✓ Urinalysis
- ✓ Wood's test or DMT plate

"Once you eliminate the impossible, whatever remains, no matter how improbable, must be the truth."

- Sherlock Holmes[iv] -

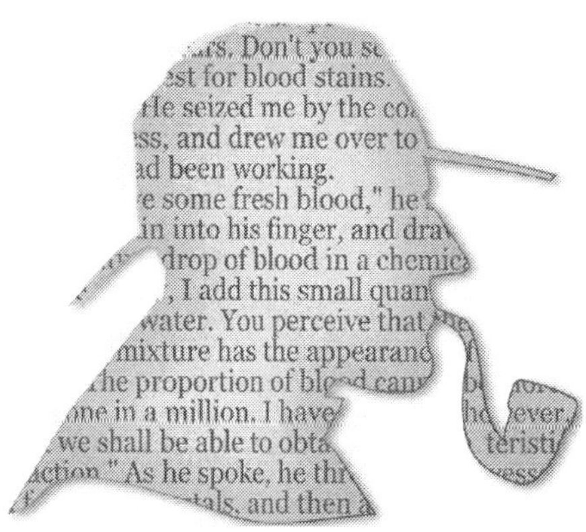

What tests do you want to run?

The tests I would choose to start with are some bloodwork **(general panel, with a pancreatic lipase and clotting times), a fecal (float and direct), and a giardia test**. I'd also recommend **abdominal radiographs** (if ol' Bernard really wants to get jiggy with the pocketbook). Since I am only partially brilliant at this stage in my life, there are other options you could have picked that would still be appropriate (like an ultrasound, a leptospirosis test, or even a parvo test).

Why clotting times? These measure the blood's ability to clot. And one of the most common types of rat poisons inhibits clotting, so that poor little critters poisoned by it bruise and bleed to death.

Bernard approves the plan (anything to get done and out of here quickly), and you soon have some diagnostic results to peruse…

Da Poop

Fecal float: not exciting at all (no signs of parasites)

Fecal direct saline exam: a quiet, sparse population of bacteria in general

Giardia test: negative

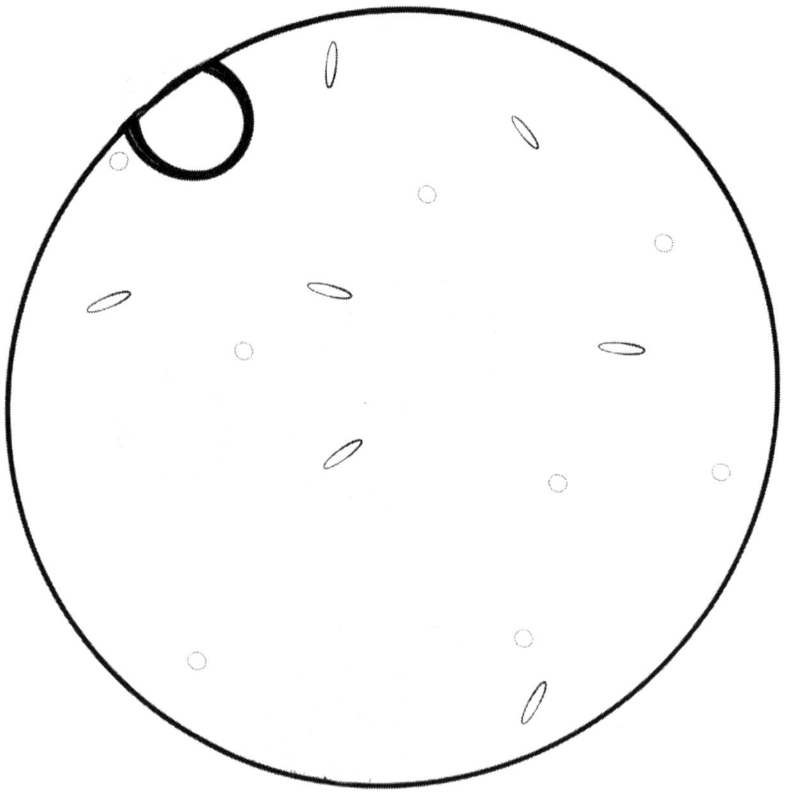

A representative field from the fecal direct saline exam of "Carl" Gallagher

Da Bloods

Client: Gallagher II, Bernard	Species: Canine	Sex: MC
Patient: Carl	Breed: Welsh Corgi	
Patient ID: 026568		

CBC

Test	Result	Flag	Reference Range[v]	
WBC	21	H	6.0-17.0	10x3/µL
Neutrophils	15	H	3-11.4	10x3/µL
Monocytes	1.89	H	0.15-1.35	10x3/µL
Lymphocytes	2		1-4.8	10x3/µL
Eosinophils	0.6		0.1-0.75	10x3/µL
Hematocrit	57	H	37-55	%
MCV	67		60-77	fL
Platelet Count	302		200-900	10x3/µL

PT and aPTT are two separate tests for blood clotting. They're normal in Carl.

PT – 10 seconds	<22 seconds
aPTT – 15 seconds	<20 seconds

Snap cPL – abnormal

cPL results are read as "normal" (negative) or "abnormal" (positive).

This is a test for pancreatitis that is run separately from the general bloodwork profile.

Client: Gallagher II, Bernard	Species: Canine	Sex: MC
Patient: Carl	Breed: Welsh Corgi	
Patient ID: 026568		

Chemistry Panel

TEST	RESULT	FLAG	REFERENCE RANGE[vi]
ALT	75	H	8.2-57 IU/L
ALP	205	H	10.6-101 IU/L
GGT	5		1.0-9.7 IU/L
tBili	0.3		0.1-0.6 mg/dL
Glucose	84		62-108 mg/dL
Cholesterol	113		116-254 mg/dL
Total Protein	7.7		5.57.5 g/dL
Albumin	2	L	2.6-4.0 g/dL
Globulin	5.5	H	2.1-3.7 g/dL
Calcium	8.2		8.7-11.8 mg/dL
BUN	66	H	8.8-26 mg/dL
Creatinine	2.5	H	0.5-1.6 mg/dL
Phosphorus	5		2.9-6.2 mg/dL
Sodium	129	L	140-154 mEq/L
Potassium	3.2	L	3.8-5.6 mEq/L
Chloride	100	L	102-117 mEq/L

DA RADIOGRAPHS

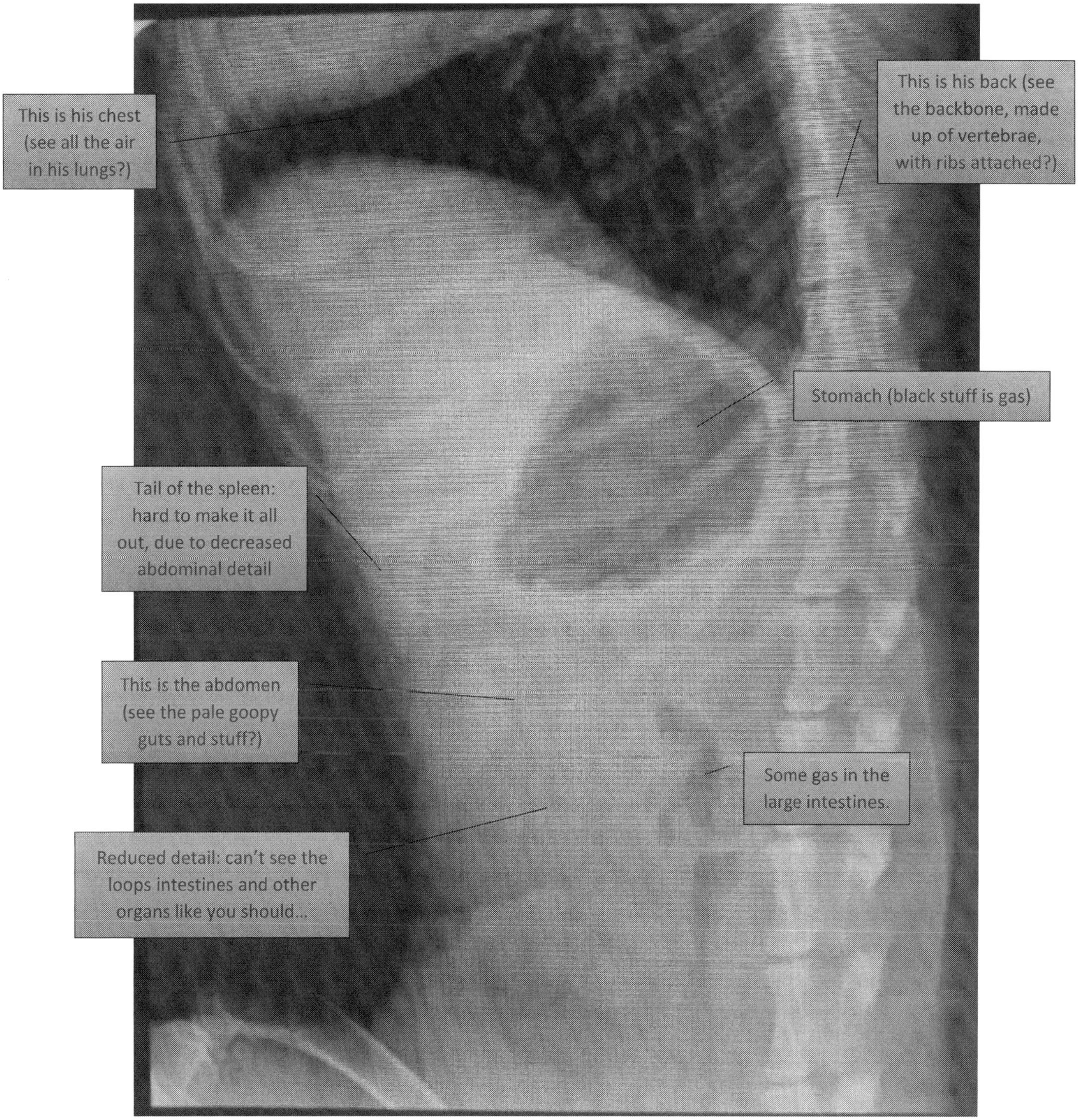

Right lateral radiograph of Carl

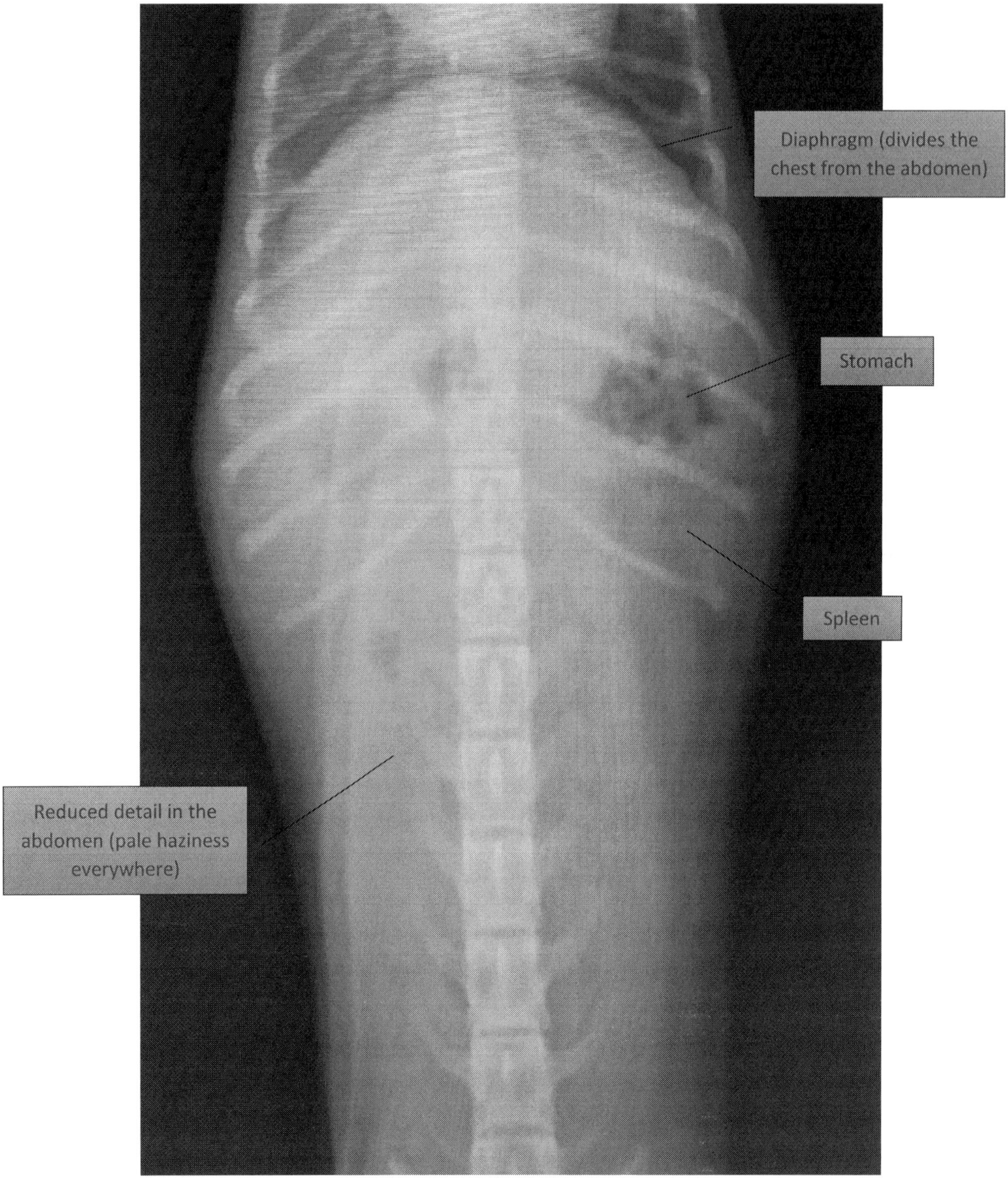

Ventrodorsal (VD) radiograph of Carl – i.e. he's on his back

You mentally take a step back and examine all the diagnostic results to try to get a picture of what's going on. What are these tests trying to tell you, Timmie? We have…

Remember! You can check the back of the book for a more detailed explanation of the information on the bloodwork panels and radiographs—and for definitions of the words!

- Scant bacterial population on fecal (with no parasites).
- Elevated white blood cell counts on bloodwork.
- Low albumin protein on bloodwork.
- Mild azotemia (increase in BUN and Cr together) on bloodwork.
- Mild elevation in some liver values on bloodwork.
- Slightly elevated globulins on bloodwork.
- Decreased electrolytes on bloodwork.
- Decreased abdominal detail on radiographs (with no identifiable cause).
- cPL test abnormal

What could have done all this? Is there a disease process that can cause all of these findings? Well, to really answer that question, you're going to have to think through differential lists of each of these findings. What things could cause each of these findings to be off? Ack! Those lists of differentials/causes come in everywhere, don't they?

Here's the list of test abnormalities for Carl, with an abbreviated differential list of options under each…

- Scant bacterial population on fecal (with no parasites)
  - Antibiotics wiped out bacteria
  - Carl's GI tract is poorly developed (bad immune system, poor diet, etc.)
  - Something went wrong with your sample
- Mild elevation in some liver values on bloodwork
  - Blockage of the bile duct
  - Infection in liver
  - Infiltrative liver disease (like cancer or fatty liver)
  - Inflammation in another organ near the liver, making it angry
- Elevated white blood cell counts on bloodwork
  - Infection
  - Inflammation without infection
- Slightly elevated globulins on bloodwork
  - Infection

- Some forms of cancer
- Low albumin protein on bloodwork
  - Liver not producing enough
  - Loss through diarrhea
- Mild azotemia (increase in BUN and Cr together) on bloodwork
  - Dehydration
  - Kidneys aren't working
- Decreased electrolytes on bloodwork
  - Hormones that control electrolyte balance are out of whack
  - Loss from vomiting/diarrhea
- Decreased abdominal detail on radiographs (with no identifiable cause)
  - Free fluid in the abdomen (like bleeding internally)
  - Inflammation in an important abdominal organ
- cPL abnormal
  - Pancreatitis
  - Angry organs near the pancreas (rubbing on it or spreading inflammation to it)

OK. There's more to explain and talk about. But you should have enough information to work this out with a reasonable degree of certainty. What's ailing poor Carl?

And your diagnosis is?

STOP!!!!!!

Make a choice, adventurous sage,

Choose an answer,

Then turn the page…

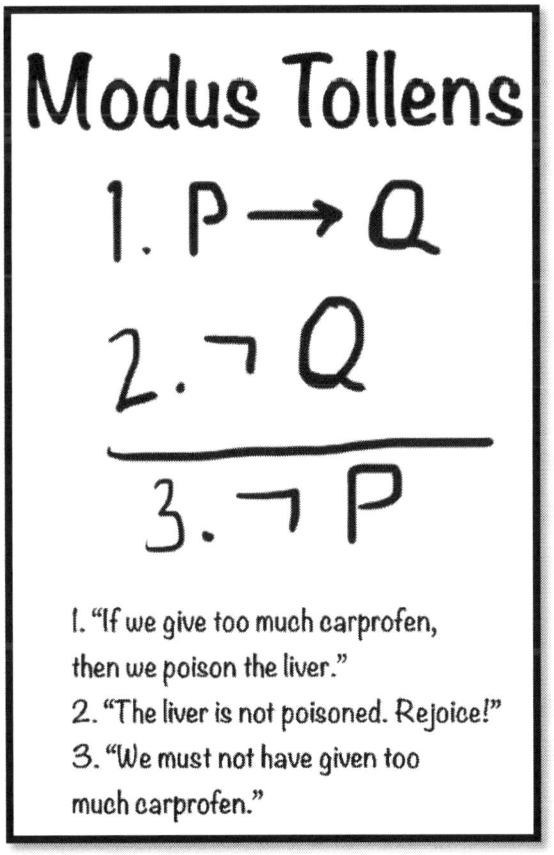

Carl is suffering from **pancreatitis**.

Is that what you picked? Good for you! You are one smart cookie! Pancreatitis is a condition where the pancreas gets all angry and inflamed (the suffix "-itis" means inflammation). Sometimes there's an infection involved and sometimes not.

Inflammation is a response of the body's immune system to fight off disease, characterized by swelling and pain. It happens mostly through the mobilization of chemicals called inflammatory mediators and white blood cells. But sometimes it can become a problem. Sometimes the inflammation gets all berserker, attacks part of the body, and makes ya sick. In Carl's case, the inflammation in his pancreas is making him sick. Pancreatitis can be acute (meaning it came on suddenly), or chronic (meaning it happens slowly over time)

You could have made arguments for other disease processes as well, so if you didn't pick pancreatitis, you're probably still a smart cookie. One reason for this is the low specificity of the snap cPL test.

Ack! Ack! Too many big words at once... Sorry. In less doctory words, that means a snap cPL test sometimes gives you a false positive result. This means a positive (or, more properly termed, "abnormal") result of a snap cPL could be from conditions other than actual pancreatitis. Things like a tumor in the front of the abdomen or a blockage near the outflow of the stomach or something inflaming the gall bladder could all conceivably cause a falsely positive snap cPL.

But...

In Carl's case, a positive cPL certainly fits the picture for pancreatitis.

Pancreatitis causes inflammation, often with increased white blood cell count (especially neutrophils). The inflammation causes the liver to be angry because it spreads to the organs sitting right next to it. Pancreatitis causes vomiting and diarrhea, which causes dehydration (which causes azotemia and electrolyte changes, and low albumin). Pancreatitis also causes decreased detail on abdominal radiographs (especially in the cranial abdomen).

"We could do an abdominal ultrasound to help confirm that this is indeed pancreatitis," you offer Bernard.

Inflammation: a response of the body's immune system to fight off disease, characterized by swelling and pain. It happens mostly through the mobilization of chemicals called inflammatory mediators and white blood cells.
Pancreas: an important little organ that makes enzymes and hormones to help digestion. He sits in the front of the abdomen, cheerfully tucked up against the first part of the small intestine near the liver and stomach. "Weird Al" Yankovic wrote a cool song about it.

"No, no," he answers, eying his watch again. "Sounds like you've got it figured out. Besides, I need to leave in time to enjoy this fantastic brie I import from France. What do we do to fix him?"

That's a good question, Bernard.

How are we gonna treat Carl? Here are some options.

a) Abdominal surgery
b) Antibiotics
c) Dewormer
d) Diuretics — Medications that take fluid out of the body.
e) IV fluids — Fluids into the veins.
f) Liver protectants
g) Nausea drugs
h) Nebulization treatment — Misting air moistened with saline (and sometimes a drug with it) over the nose to help with breathing.
i) Pain medications
j) Silver sulfadiazine — A splendid cream for wounds and burns.
k) Steroids — A class of chemical (hormones or sterols) that inhibits inflammation and enhance metabolism.
l) Subcutaneous fluids — Fluids under the skin.
m) Tomatoes — A round fruit grown in some gardens, and quite tasty on burgers.
n) Yunnan Baiyao — A Chinese herb that helps stop bleeding.

There are so many useful tools in our toolbox of healing!

What's your treatment plan? (Remember, you usually pick more than one option.)

You decide to recommend **b, e, g, and i.**

IV fluids are the mainstay of treatment for pancreatitis. IV fluids need to be administered over time through an IV pump, so Carl will have to stay in the hospital while he gets them.

In this case, IV fluids are better than subcutaneous fluids. Carl is so very dehydrated that IV fluids (which are more effectively given and controlled than SQ fluids) are called for. Also, because Carl is so very sick, they will help support the blood pressure and treat shock. Subcutaneous fluids are generally used for mild to moderate dehydration, where hospitalization just isn't practical. He should be on fluids until he's rehydrated and feeling better. This is often a course of 2-3 days in dogs with pancreatitis.

Drugs: you have shelves full of them at your disposal! You decide to put Carl on some **antibiotics**. This is because his white cell counts and globulins are high. You also give him several injections for **nausea** and vomiting ("antiemetics"). Pancreatitis is usually painful, and Carl is acting painfully today when you palpated his abdomen, so you start him on some **pain injections** too. At one time, the administration of steroids to reduce inflammation in the pancreas was a common practice. But now, because their anti-inflammatory properties are not specific and they have a lot of side effects, we've realized steroids are usually not helpful for this.

You explain your plan to Bernard, and your technicians go over an expensive estimate for several days of hospitalized and supportive care. Bernard is so happy to finally rush out of the clinic to his cheese tasting and book club that he barely grumbles as he pays his bill.

You scribbled out some brilliant math to calculate fluid rates and drug dosages and leave your newly hospitalized patient in the expert hands of your technicians. There are more vaccines to give!

> "Genius is fine for the ignition spark, but to get there someone has to see that the radiator doesn't leak and no tire is flat."
>
> —Rex Stout (*The Doorbell Rang,* Nero Wolfe)[vii]

Case #3

"There's a surprise for you in the next room, Doc," Fred says ominously after you finish polishing up your notes on a dachshund with allergies. Hm, OK... You open the door with understandable caution.

Your first impulse on seeing the thing licking its paws on the exam table is to yell, "What the hot-buttered monkey is that wrinkly thing? Where'd all its hair go? Why're its eyes so big?"

But you exert an impressive amount of self-control and look thoroughly nonplussed on the outside. Experience has taught you that anything this ugly must be an expensive purebred and, being the logical person you are, you reason expensive purebreds tend to be owned by sensitive people with lots of money. And that's two good reasons to be careful what you say about the wrinkly monster purring happily and flicking its naked tail in front of you.

A smiling young lady named Summer introduces herself. "And this is Mole Rat, my Sphinx cat."

You cast a dubious eye at the beast before you and try not to recoil as you rub your hands over his hairless body. He pushes his head into your hand. He's surprisingly velvety, and you reluctantly relax just a bit. Mole Rat, it turns out, may not be some kind of mutant space monster after all, but a sweet kitty.

"What's going on with this guy today?"

"Well, he's just not been acting himself for the past two days. He has been hiding a lot and seems lethargic."

"Eating and drinking?" You remember asking this question earlier today and wish you'd rephrased it. But Summer is less literal than the bookish Mr. Gallagher II.

"A little bit, but not nearly as much as he used to."

"Is he vomiting or having diarrhea?"

Lethargic: lacking energy and acting listless.

"No."

"Have you noticed any coughing or sneezing?"

"No."

"Is he an indoor or outdoor cat?"

"Oh, he's an indoor cat. But… He does get outside sometimes, mostly when I get sushi and turn on one of my historical docudramas. That cat really hates sushi! I thought all cats loved fish."

You feel a bit more kinship toward the strange creature (who steps gingerly up to you and licks your hand with his scratchy little tongue). "You like historical docudramas? I watched a ton of those last year. Now I'm more into observational comedy sitcoms."

"Lit, man! I just finished *Pistachios for Comanches* on blu-ray… What a great series!"

"Truth. It wasn't half bad…wait…you have a blu-ray copy already? Didn't *Pistachios for Comanches* just air?"

She blushes a little. "Yeah, well, my cousin gave it to… Er, did you see the part where Black Horse did that war dance?"

"Oh yeah! That was high art, wasn't it…" You nearly get distracted talking about your favorite shows with this pretty client, but Mole Rat nibbles your finger with his unnaturally small teeth to remind you you're here to be his doctor. You force your brain back to business and move on to do an exam.

Step 1 – Organize the History

Signalment:

- Male neutered
- Sphinx cat
- 4 years old

Presenting Complaints:

- Lethargy
- Decreased appetite
- Hiding

Step 2 – Gather Clues

Physical Exam:

Abdominal palpation: WNL.

Attitude: Quiet, slightly lethargic, but responsive.

Cardiac: WNL. No murmur or arrhythmia noted.

Ears: WNL. Canals appear clear.

Eyes: WNL.

Integument: Normal skin turgor. Weird, velvety skin with no fur (breed-related). It's oily too, and full of folds and swells… It's kinda gross and you move on from the integument part of your exam quickly so you won't have to feel it anymore.

Lymph Nodes: WNL.

Musculoskeletal: WNL.

Neurologic: No significant findings.

Oronasal: No significant findings.

Pulmonary: The patient is eupneic, with normal rate and effort to his respiration and no coughing or sneezing noted.

Urogenital: WNL. Neutered male.

Vitals: MM pink and tacky. CRT<2 seconds. Pyrexia noted: 104 F rectal temperature.

"Tacky" = sticky gums. A sign that he is at least slightly dehydrated.

Differential Diagnoses (the possible diseases):

Abscess

Cholangiohepatitis

Heat stress

Hepatic dysfunction

Hypersensitivity reaction

Gastroenteritis

Infection

Metabolic disease (like diabetes or hyperthyroidism)

Pancreatitis

Renal failure

Toxicity

Abscess: infection trapped under the skin, forming a pus bubble.

Cholangeohepetitis: a disease where the bile duct and surrounding tissues in the liver and gallbladder are all inflamed.

Eupneic: a fancy word for breathing normally. The opposite of dyspneic (breathing abnormally).

Pyrexia: a fancy word for having a fever.

Decision time!

What tests do you want to run first?

Here's what you can do in your clinic:

- ✓ Basic bloodwork (a chemistry panel and CBC)
- ✓ Blood gas analysis
- ✓ Clotting times
- ✓ Cytology
- ✓ Fecal floatation and direct saline exam
- ✓ FeLV/FIV test
- ✓ Giardia test
- ✓ Heartworm test (with rickettsial diseases)
- ✓ Leptospirosis test
- ✓ Pancreatic lipase (fPL or cPL)
- ✓ Parvo test
- ✓ Pregnancy test
- ✓ Radiographs (X-rays)
- ✓ Skin scraping
- ✓ Thyroid hormone level (T4)
- ✓ Ultrasound
- ✓ Urinalysis
- ✓ Wood's test or DMT plate

Two different viruses cats get from fighting. They attack the immune system and open the way to other infections.

"'Never mind,'
I said crisply.
'I have my methods.'"
- P.G. Wodehouse[viii] -

Wait!

Before you answer, think carefully through our clues from the physical exam, history, and signalment. If we pay attention to these clues, we can often prioritize our differential list (i.e., guess which causes of the symptoms are most likely) even before we run tests. Then we can strategically pick tests that make more sense.

Mole Rat is lethargic, eating less, has a fever, and is slightly dehydrated. I would rank differentials that cause fever (like infection, abscess, hypersensitivity, heat stress, or maybe pancreatitis) higher, and prioritize tests that investigate those disease processes.

What tests do you want to run?

Tests to investigate a fever usually start with **bloodwork**, so let's do that! Running radiographs or an ultrasound would not make much sense just at this point.

I would also do a **FeLV/FIV test** and maybe a **snap fPL** (just for fun)...

Snap fPL: normal

FelV/FIV: negative/negative

Client: Summer Q. Smiley	Species: Feline	Sex: MC
Patient: Mole Rat	Breed: Sphinx	
Patient ID: 023573		

CBC

TEST	RESULT	FLAG	REFERENCE RANGE[ix]	
WBC	35	H	5.5-19.5	10x3/mm^3
Neutrophils	19	H	2.5-12.5	10x3/mm^3
Monocytes	0.7		0-0.85	10x3/mm^3
Lymphocytes	2		1.5-7	10x3/mm^3
Eosinophils	1		0-1.5	10x3/mm^3
Hematocrit	30		30-45	%
MCV	50		41.0-54	fL
Platelet Count	250		150-600	10x3/mm^3

Client: Summer Q. Smiley	Species: Feline	Sex: MC
Patient: Mole Rat	Breed: Sphinx	
Patient ID: 023573		

Chemistry Panel

TEST	RESULT	FLAG	REFERENCE RANGE[x]	
ALT	30		8.3-35	IU/L
ALP	55		12-65	IU/L
GGT	2		1.8-12	IU/L
tBili	0.4		0.1-0.5	mg/dL
Glucose	164	H	61-124	mg/dL
Cholesterol	91		71-161	mg/dL
Total Protein	9.0	H	5.7-8.0	g/dL
Albumin	2.6		2.4-3.8	g/dL
Globulin	6.4	H	2.4-4.7	g/dL
Calcium	2		2.0-2.7	mg/dL
BUN	20		15-31	mg/dL
Creatinine	1.1		0.5-1.9	mg/dL
Phosphorus	5.5		4.0-7.3	mg/dL
Sodium	149		146-159	mEq/L
Potassium	4.5		3.8-5.3	mEq/L
Chloride	110		108-130	mEq/L

That's it then… You ponder the clues your diagnostics gave you. The test for pancreatitis and feline leukemia (FeLV) and FIV were both negative, so it's not those (assuming the tests are not falsely negative, which is not likely in this case).

Where's this fever coming from? Are the bloodwork results more consistent with a non-inflammatory fever (from things like hypersensitivity reaction or heat stress), or an inflammatory one (from things like an infection or an abscess)?

Hm… An abscess…

You obviously know what an abscess is, but just in case you forgot, I'll remind you. An abscess is a pocket of infection that builds up pus trapped under the skin, which causes fever, among other things. It just gets more swollen and painful and infected as time progresses, until the pressure eventually causes it to rupture. Cats commonly get abscesses from fighting with other cats. And didn't Summer say that Mole Rat goes outside sometimes?

You reluctantly decide it's a good idea to do another check of Mole Rat's skin. You put on double pairs of gloves and go for it, carefully running your hands back over the purring little kitty again and paying more attention to all those weird ridges and wrinkles to make sure they're all supposed to be there. This time, when you get to his flanks, near the base of the tail, you notice a pronounced swelling that you thought before was just another strange ridge upon his skin. He twitches when you touch it and hisses a little bit.

"Well now, here's a thing…" you murmur.

"What thing? What thing?!?!?" Summer bounces out of her seat like a curious golden retriever, and you realize you should have kept your mouth closed.

"I think this could be an abscess. It's swollen and hurts Mole Rat when I touch it. Let's poke it with a needle!"

A handy topical anesthetic (drugs that cause sleep or numbness) in a gel form.

She gives you a look.

"After I numb it with a bit of lidocaine gel, of course."

You put the gel on and wait around 10 minutes (during which you rush off to give a pit bull puppy a vaccine). Then you poke that swelling with a needle. Sure enough! From inside the swelling, you draw up a light brown liquid. You have Fred schmear some on a slide and stain it to do a **cytology**.

This is a test you decide to run as circumstances changed. Sometimes you must make a choice without a lot of time to sit and think about it. Remember, cytology is looking at cells smeared on a slide with a microscope.

Anesthetic: a drug that causes sleepiness (like for doing surgery) or numbness.
Topical: to put on the surface of the skin (as opposed to giving it as a pill or an injection).

You peek into the microscope to find a busy bunch of cells partying.

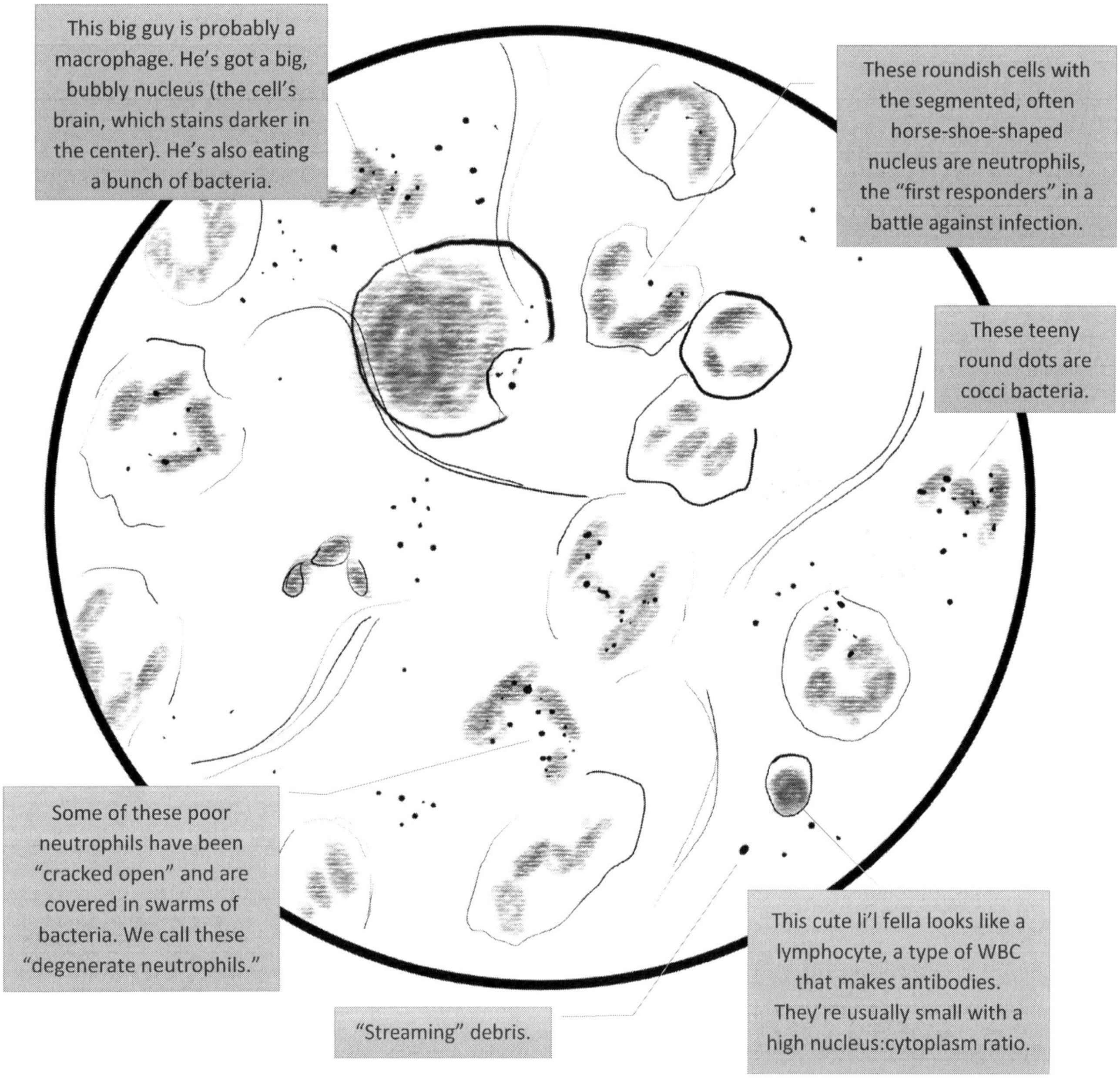

Cytoplasm: the "blood" of a cell—the central gooey stuff that the organelles float around in, surrounded by the membrane (lining) of the cell.

Macrophage: a type of white blood cell whose job is mainly to engulf and eat up invading bacteria and things.

Nucleus: one of a cell's most important "organelles" (teeny organs), the nucleus is the brain of the cell and is where the DNA or RNA is stored. It stains darker than the cytoplasm on cytology.

Nucleus:cytoplasm ratio: comparison of how bit the nucleus is to the amount of cytoplasm. This is one of many ways to help describe a cell under a microscope. A high N:C ratio = a bigger, fatter nucleus.

And your diagnosis is?

____

STOP!!!!!!

Don't turn the page until you've made a diagnosis…

Cross-Section of a Typical Eukaryotic Cell

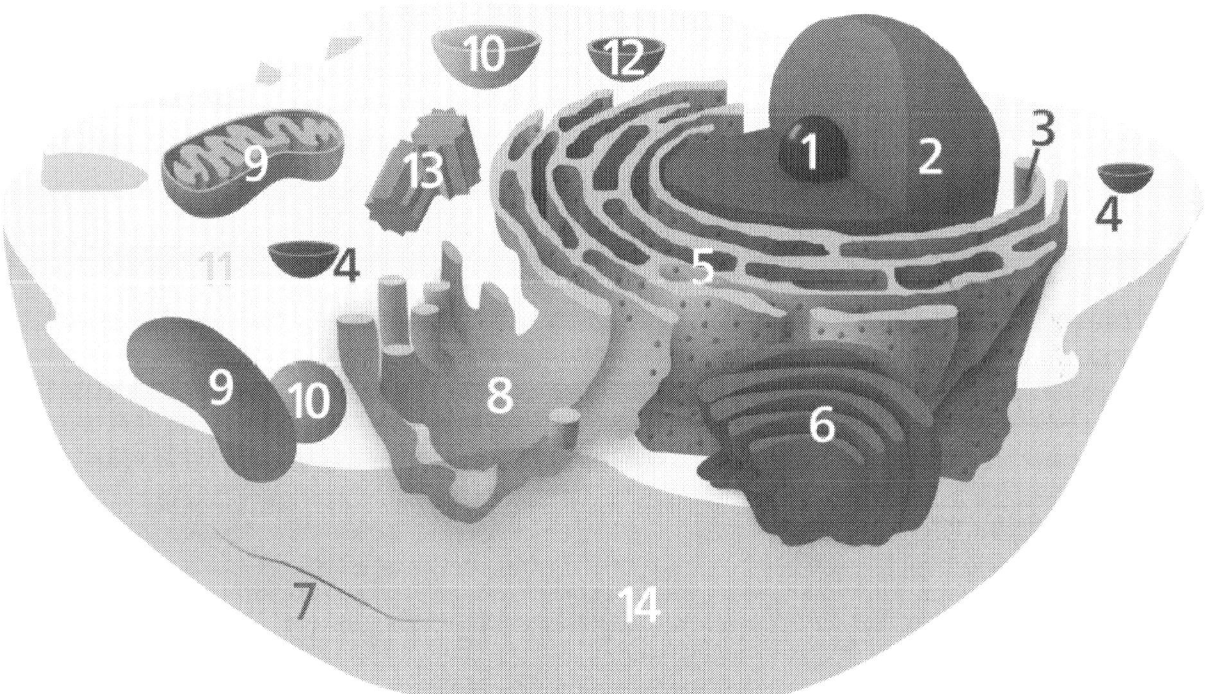

1. Nucleolus
2. Nucleus
3. Ribosomes (dots on the RER)
4. Vesicle (carries stuff through cell)
5. Rough Endoplasmic Reticulum (RER)
6. Golgi apparatus
7. Cytoskeleton (series of tubes that give cell its shape)
8. Smooth endoplasmic reticulum
9. Mitochondrion (makes energy for cell)
10. Vacuole
11. Cytoplasm
12. Lysosome
13. Centriole
14. Cell membrane (its "skin")

"Gross!" you exclaim. "These cells are having an infection party!"

This doesn't win you any points with Ms. Summer, but it was fun to say.

And yes, you're right. What you're seeing under the microscope is basically sheets of white blood cells (especially neutrophils) chomping and hammering down on hordes of bacteria. This is a "suppurative exudate," a polite and medical way to say "pus."

You've just diagnosed an **abscess**! That's what pus looks like under the microscope. Wait till you see it outside the lens. Things are about to get much, much grosser…

A cat abscess usually comes about from the skin being poked by another cat's claws, then getting infected underneath. Mole Rat was probably running away from a fight, like a dirty coward, since his abscess is right above his tail.

What do you want to do to treat this poor naked cat-beast?

Pick from these options… (Hint: more than one!)

a) Antibiotics

b) Buy him a jacket

c) Chest coupage

> Slapping the chest gently to break up gunk in the lungs.

d) Diet change

e) Drugs for nausea

f) IV fluids

g) Lance the abscess

h) Pain medications

i) Run and hide

j) Steroids

k) Subcutaneous fluids

l) Surgical skin graft

m) Warm compresses

n) Sedation

STOP!!!!!!

Don't turn the page until you've picked some treatments…

I would plan on **a, g (with some n), and h**. (Agh!)

You need to sedate Mole Rat and **lance that abscess**! This will allow it to drain and relieve the pressure, and you can clean out as much gunk as possible to give him a good head start in healing from the infection. I'd also start him on some **antibiotics** and **pain medication**.

There are some cases (especially if cost is a concern or the patient can't be sedated for medical reasons) when an abscess could be treated more conservatively with antibiotics and warm compresses once daily. The compresses will thin the skin over the abscess as it swells more until it ruptures all over the owner at home and relieves the pressure on its own. Summer wisely declines the plan of warm compresses and is happy to have you lance and clean the abscess at the clinic.

Not too many minutes later, you've given this wrinkly, naked little hob-goblin a pain injection and sedative. He feels pretty good and is rolling around half drunk on your wet table (an exam table with a sink and drain). You wear more gloves, screw up your courage, and take a scalpel blade in hand. With a deft flick o' the fingers, you poke the blade into the swelling above Mole Rat's tail and stand back. Nothing happens. You get a curved tip syringe with some dilute betadine and start poking it into the hole you've made to flush it. There's a little popping feel as you jab through some fibrin and GLUCK! A stream of revolting, brown-colored, death-smelling pus streams out of the swelling.

Yes! That's more like it! Very satisfying. Like watching meat gravy poured over your turkey on Thanksgiving...[3]

After you finish trying not to vomit, you flush the lanced abscess and clean the surrounding skin. Then you leave your patient in the capable hands of your technicians to recover from his sedation before going home.

You've fixed an ailing Sphinx cat! Great job! Give yourself a high five!

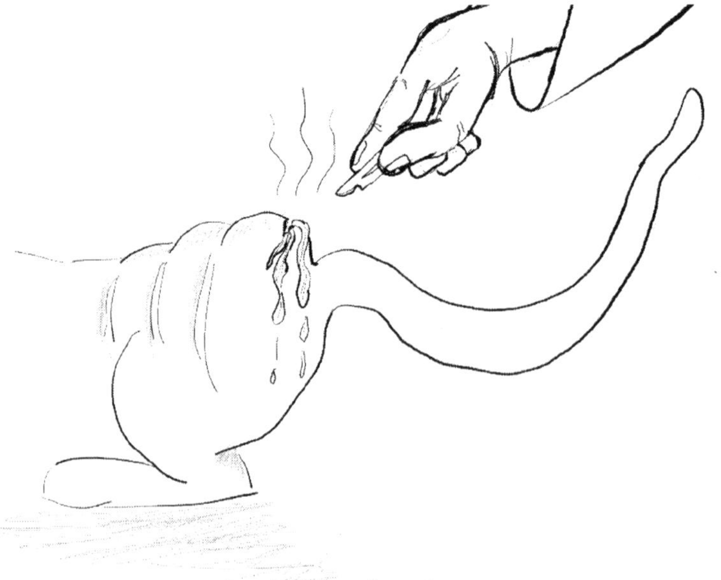

[3] Apologies to any squeamish readers. We veterinarians always describe the most disgusting things we see in terms of food items. It's sort of a guild tradition.

Case #4

That last case made you hungry. You're about to leave the clinic for lunch when a car comes roaring into a parking spot and a panicked guy jumps out and runs to the back seat. That's never a good sign. He's a tall man, with lanky, tattooed arms displayed prominently outside his sleeveless shirt. There's a big, gold chain around his neck, and a bandana tied around his head.

You try never to read a book by its cover, but in this case, you know the initial impression of a wannabe "cholo"[4] gansta is very accurate. You've met this man before. His name is Arturo Melendez, and he's the son of one of your clinic's oldest clients. You also know that, though he embraces a gangster style, Arturo did not grow up in a streetwise life: rather, he's got a master's degree in chemical engineering, and his parents are both wealthy entrepreneurs. You should go back to not reading a book by its cover. Arturo interrupts these reflections by digging into the back seat of his car and hauling out a big, bleeding pit bull.

"*Ayudame!*[5] You gotta help fix Asesino!" You've been living in New Mexico long enough to know that *Asesino* is Spanish for "Murderer," and the dog in Arturo's arms appropriately tries to bite your arm off. Even so, you can't let him bleed out in the parking lot. What kind of doctor would you be if you treated only the dogs that didn't want to eat you?

"Come inside, let's take a look."

"What's up Doc?" Fred asks as Arturo charges through the door with poor Asesino. Your tech thinks he's being funny, but you don't have time just now to notice. From your safe distance, you can see a gaping wound on the left side of Asesino's neck, and several deep-looking puncture wounds on his chest. There are multiple smaller wounds on his shoulders and neck and the top of his head as well. He's breathing rapidly but without increased effort that you can tell. He snarls at you from his owner's arms, and you can see his gums are still pink. That's the best you can do for an initial exam.

"What happened to him?" you ask.

---

[4] A cholo is a distinctive type of Hispanic gangster style. The style is associated with a specific outfit (a sleeveless shirt with a bandana tied low over the forehead, baggy khaki pants, and a loose flannel shirt), low-rider cars, and specific lifestyle preferences.

[5] "Help me!"

"Oh! Um…" Poor Arturo is covered in blood already. "He was attacked…by a coyote…or maybe a wolf. It was one big hombre!"

You have the man stand with his dog on the scale. Then subtract what he says he weighs to get a weight on this painful, aggressive critter.

It's an emergency! What are you going to do next?

Your new patient needs help. Also, every second you stand there is another bite out of your lunch hour! What is your plan of attack?

There's a practical first step you need to do before anything else can be done to this vicious dog…

Do something!

Do something!

Hurry!

AAAAAAAAGH!

No, no…

Take a breath. One of the best things to do in an emergency situation is to stop panicking, calm down, and think of the very next step.

That next step is… ?

"Right is Right even if nobody does it. Wrong is wrong even if everybody is wrong about it."

—G. K. Chesterton[xi]

If you decided on a **sedative** first, great work! That will help your patient and keep your staff and client from getting chomped on. A **pain injection** is needed too. Then you can breathe for a minute, get a more thorough exam done, and plan some tests and treatments to take care of Asesino from there.

The skill of keeping a cool head—standing back and taking a calm look to decide what you should do first in an emergency—is key to success in a case like this.

You pop an IM injection into the poor bleeding, raging pit bull and let him calm down in an oxygen cage (since he has puncture wounds along his chest, and you're not sure of their extent yet), while your technician learns more history. Arturo doesn't add much more to the story. His dog has been healthy prior to today. He takes a steroid for allergies and an iron supplement, but no other medication. There's been no coughing, sneezing, vomiting, or diarrhea noted. He was outside today, running on the owner's property outside of town on a few acres, and was attacked by a wild canid (you assume it was a coyote since wolves are super rare in the Santa Fe area). Then Arturo rushed him to the clinic.

OK. Time to do a full exam and organize your thoughts…

Step 1 – Organize the History

Signalment:

2 year old

Male intact

American Staffordshire Terrier ("Pit Bull")

Presenting Complaints:

Attacked by a wild animal

Wounds

Step 2 – Gather Clues

> Looks like you're learning to be more methodical with your PE. Good job!

Physical Exam:

Abdominal palpation: WNL.

Attitude: sedate.

Cardiac: heart sounds muffled. No murmur or arrhythmia noted.

Canid: a member of the dog species. (Canine is the adjective form of the noun canid.)
IM: intramuscular. A shot into the muscles.

Ears: WNL. Canals appear clean.

Eyes: WNL.

> It's generally helpful to be specific when describing a wound, so your "word picture" will help other medical professionals understand what the wound looked like at first if they need to consult the records later.

Integument: long laceration (~3") on the left thoracic region, extending into the subcutis with muscle visible. Extensive pocketing is noted, and some muscle damage is noted. No "sucking" or "blowing" motion is noted in the wound. Puncture wounds in the neck, head, and shoulder area. Smaller lacerations on the neck, head, and shoulder area. Some are older wounds, and some have a suppurative exudate and foul smell. Multiple scars are visible on the head and shoulders.

> These could be signs of a puncture into the chest cavity.

Lymph Nodes: unremarkable.

Musculoskeletal: No fractures or luxations palpated. Flexion and extension of peripheral joints WNL.

Oronasal: WNL.

Pulmonary: Lungs sound clear.

Urogenital: Unremarkable. A urinary bladder is palpable with normal tone and size.

Vitals: MM pale pink, CRT<2 seconds.

Differential Diagnoses:

Penetrating thoracic trauma

> Something (like a wound or an arrow or a 16-foot-long Swiss pike) poking through the chest wall.

Pneumothorax

Pulmonary contusions

Soft tissue injury (lacerations, muscle damage, ruptured ligaments, etc.)

Subluxation or fractures (like chip fractures, or small fractures you can't palpate)

Traumatic brain injury

Visceral damage

Luxation: being out of joint.
Peripheral: outside of the brain or spinal cord. Then means limbs (when talking about the musculoskeletal system).
Pneumothorax: air trapped between the lungs and body wall. This is something often seen from punctures to the thorax.
Pocketing: in a wound, this is a term for an empty space formed by the skin pulling away from the body.
Pulmonary contusions: bruises (and bleeding) in the lungs.
Subcutis: the layer of fatty tissue just under the skin (helps insulate the body and bring blood to the skin, among other things).
Subluxation: a partial luxation.
TBI: traumatic brain injury.
Thoracic: related to the chest.
Visceral: related to the soft organs in the abdomen.

Decision time!

What tests do you want to run first?

Here's what you can do in your clinic:

- ✓ Basic bloodwork (a chemistry panel and CBC)
- ✓ Blood gas analysis
- ✓ Clotting times
- ✓ Cytology
- ✓ Fecal floatation and direct saline exam
- ✓ FeLV/FIV test
- ✓ Giardia test
- ✓ Heartworm test (with rickettsial diseases)
- ✓ Leptospirosis test
- ✓ Pancreatic lipase (fPL or cPL)
- ✓ Parvo test
- ✓ Pregnancy test
- ✓ Radiographs (X-rays)
- ✓ Skin scraping
- ✓ Thyroid hormone level (T4)
- ✓ Ultrasound
- ✓ Urinalysis
- ✓ Wood's test or DMT plate

What tests do you want to run?

Run tests, run!

Radiographs or ultrasound (or maybe both). Imaging the thorax, in particular, might be a good idea in this dog, since his wounds are over his front half. Arturo also approves a **general blood panel**.

You obtain two radiographs. Take your time to look at them and see if you can identify some of the normal landmarks before you try to decide what's not normal.

Fill in the boxes.

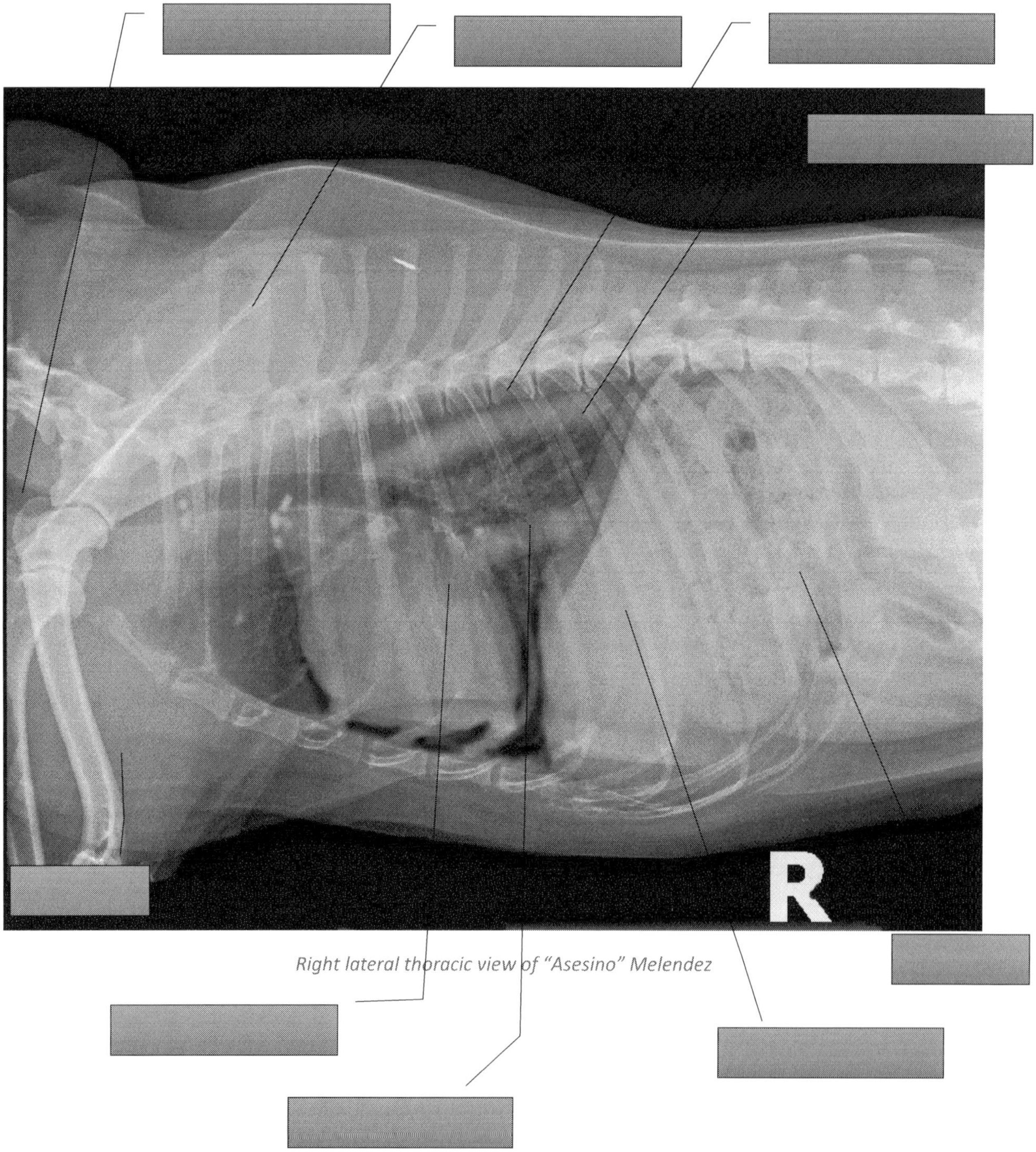

Right lateral thoracic view of "Asesino" Melendez

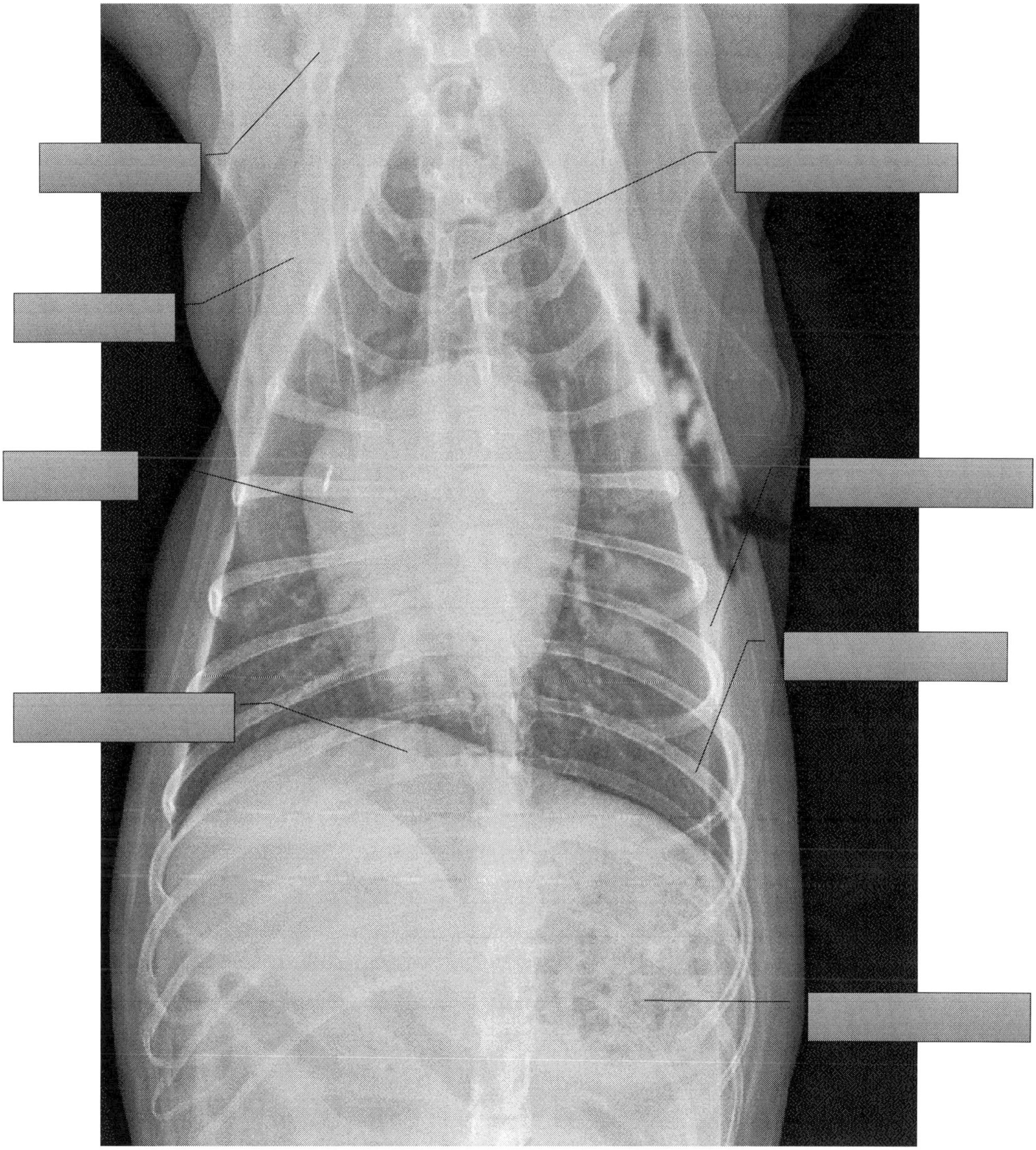

VD thoracic view of "Asesino" Melendez

Lateral: looking from the side.

VD: ventrodorsal; looking from the chest toward the spine (patient laying on back).

OK. Here are the answers. How'd you do?[6] The black stuff surrounding the heart is air in the lungs…

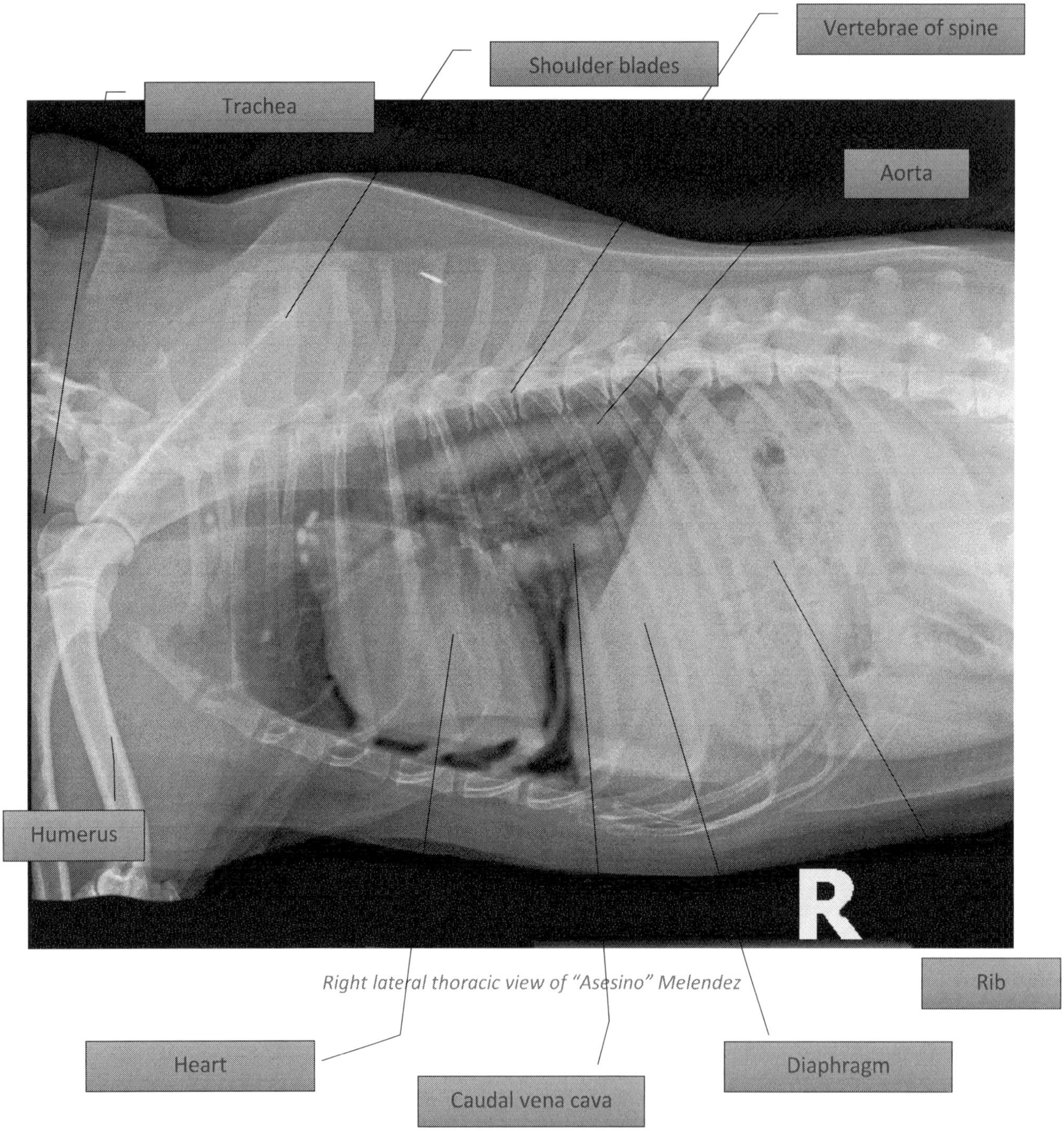

Right lateral thoracic view of "Asesino" Melendez

[6] Give yourself a high five and 3—no, 3.67—points for each right answer!

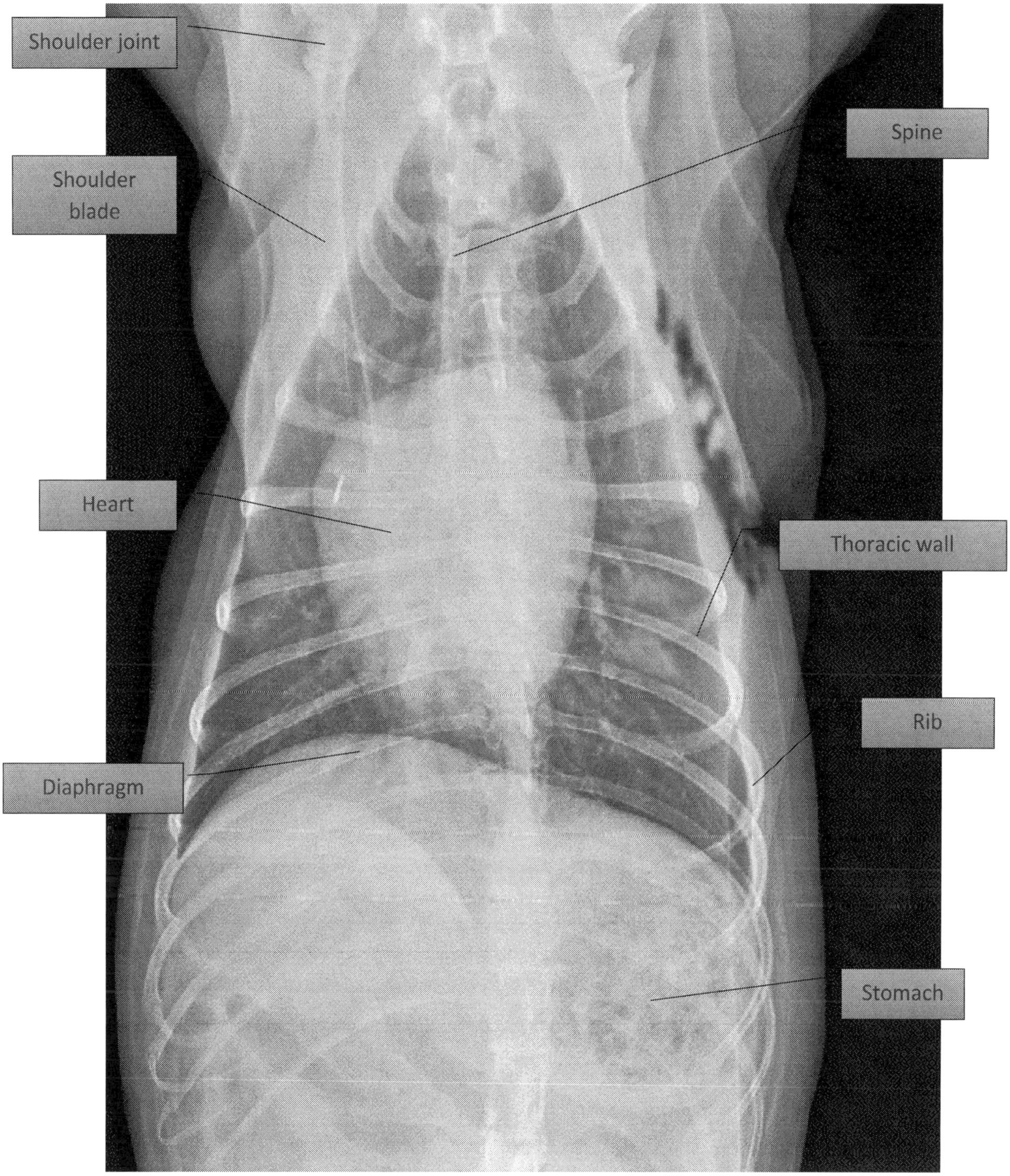

VD thoracic view of "Asesino" Melendez

Now you've got your bearings with some normal radiographic anatomy…

Do you see anything abnormal on these radiographs?

Look back over the past few pages before you move on from this one…

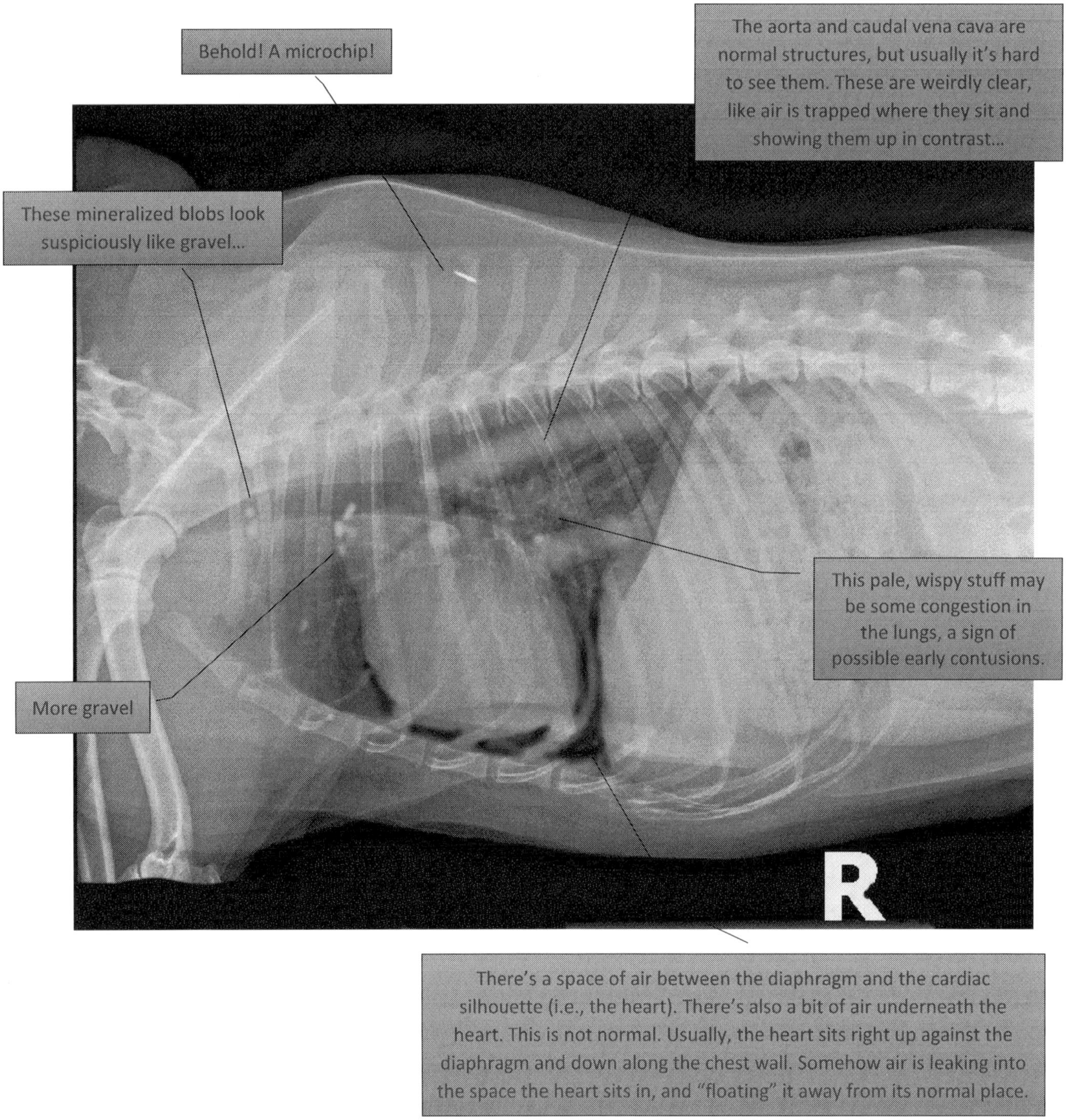
Behold! A microchip!
The aorta and caudal vena cava are normal structures, but usually it's hard to see them. These are weirdly clear, like air is trapped where they sit and showing them up in contrast…
These mineralized blobs look suspiciously like gravel…
This pale, wispy stuff may be some congestion in the lungs, a sign of possible early contusions.
More gravel
R
There's a space of air between the diaphragm and the cardiac silhouette (i.e., the heart). There's also a bit of air underneath the heart. This is not normal. Usually, the heart sits right up against the diaphragm and down along the chest wall. Somehow air is leaking into the space the heart sits in, and "floating" it away from its normal place.

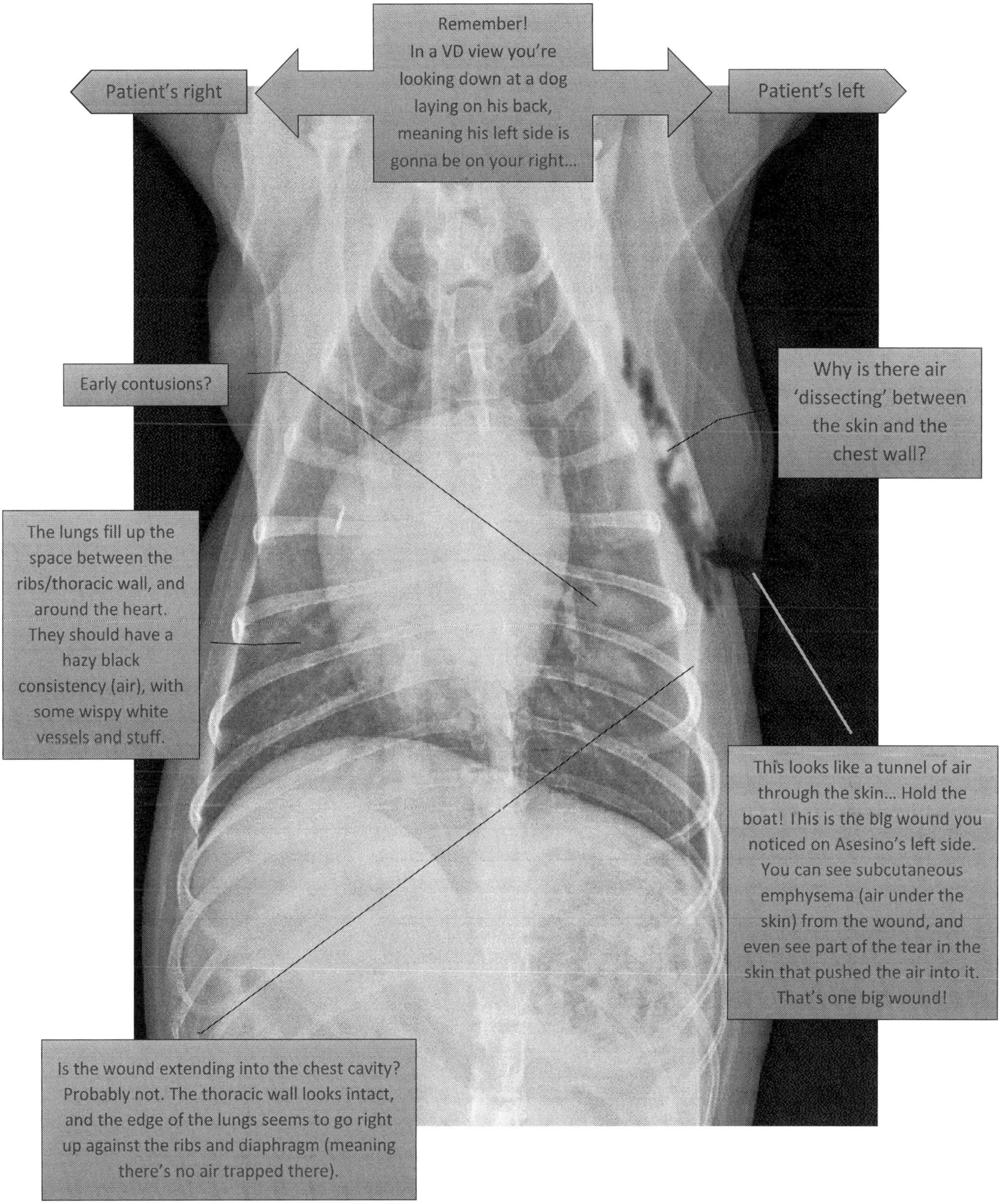
Remember!
In a VD view you're looking down at a dog laying on his back, meaning his left side is gonna be on your right...
Patient's right
Patient's left
Early contusions?
Why is there air 'dissecting' between the skin and the chest wall?
The lungs fill up the space between the ribs/thoracic wall, and around the heart. They should have a hazy black consistency (air), with some wispy white vessels and stuff.
This looks like a tunnel of air through the skin... Hold the boat! This is the big wound you noticed on Asesino's left side. You can see subcutaneous emphysema (air under the skin) from the wound, and even see part of the tear in the skin that pushed the air into it. That's one big wound!
Is the wound extending into the chest cavity? Probably not. The thoracic wall looks intact, and the edge of the lungs seems to go right up against the ribs and diaphragm (meaning there's no air trapped there).

As you're pondering these radiographs in your heart, your lovely assistant Bojka (who immigrated from Serbia) hands you some bloodwork results. You almost forgot bloodwork was pending!

Let's take a look, shall we?

Client: Melendez, Arturo	Species: Canine	Sex: M
Patient: Asesino	Breed: Am. Staff. Terrier	
Patient ID: 026569		

CBC

Test	Result	Flag	Reference Range[xii]	
WBC	19	H	6.0-17.0	10x3/µL
Neutrophils	14.1	H	3-11.4	10x3/µL
Monocytes	1.71	H	0.15-1.35	10x3/µL
Lymphocytes	3		1-4.8	10x3/µL
Eosinophils	0.6		0.1-0.75	10x3/µL

Elevated WBC = inflammation, probably infection with this combination of elevated counts. Where could this be coming from?

Hematocrit	35	L	37-55	%
MCV	74		60-77	fL
Platelet Count	302		200-900	10x3/µL

There doth appear to be a mild anemia. Where is that coming from?

Client: Melendez, Arturo	Species: Canine	Sex: M
Patient: Asesino	Breed: Am. Staff. Terrier	
Patient ID: 026569		

Chemistry Panel

TEST	RESULT	FLAG	REFERENCE RANGE[xiii]	
ALT	50		8.2-57	IU/L
ALP	25		10.6-101	IU/L
GGT	5		1.0-9.7	IU/L
tBili	0.3		0.1-0.6	mg/dL
Glucose	84		62-108	mg/dL
Cholesterol	119		116-254	mg/dL
Total Protein	4	L	5.5-7.5	g/dL
Albumin	3		2.6-4.0	g/dL
Globulin	5.5	H	2.1-3.7	g/dL
Calcium	8.8		8.7-11.8	mg/dL
BUN	24		8.8-26	mg/dL
Creatinine	1.1		0.5-1.6	mg/dL
Phosphorus	5		2.9-6.2	mg/dL
Sodium	149		140-154	mEq/L
Potassium	4.1		3.8-5.6	mEq/L
Chloride	115		102-117	mEq/L

There are more proteins in the blood than just albumin and globulin. TP will go down with external blood loss. That would make sense of the anemia as well.

Elevated globulin = infection

STOP!

What do you think of your clues so far? Take a step back, look at the entire picture, and come up with an assessment of Asesino's condition… **AND YOUR DIAGNOSIS IS??**

You're right! Those radiographs show us a big ol' wound extending down to the chest wall but not through it. This is **soft tissue trauma** (something hath mangled her skin and muscles). There's gravel pushed into multiple wounds, visible on the lateral view (remember, a radiograph is a one-dimensional representation of a three-dimensional animal: the gravel that looks like it's inside him is just inside the wounds on his skin). The lung edges seem to go right up to the body wall on the lateral view, so there *probably* isn't penetration into the chest.

BUT we seem to have air floating around the heart, the aorta, and vena cava. How can that be? Where do those three structures sit? Maybe, just maybe there's a less obvious second problem here you didn't even put on your differential list before. Uh oh!

Because the scorching fires of vet school hath forged you into a brilliant veterinary anatomist, you know that the heart, the caudal vena cava, and the aorta (along with some other important stuff) are contained within a membrane called the mediastinum. And further, you know that in dogs, the mediastinum is completely separate from the rest of the chest. There is air inside the mediastinum in this patient.

Asesino is suffering from **pneumomediastinum** (in addition to her wounds). Do you know another super important structure connecting with the mediastinum? The trachea ("windpipe"). Yipe! Could this patient have a tear in his trachea? That sounds bad.

What about the bloodwork? You see signs of inflammation (high white blood cell counts and globulins), probably from an infection. You see several older wounds on Asesino that look puffy and ooze pus, so that's not too surprising. There's also evidence of blood loss (anemia with a drop in total protein), but that's not surprising since your patient arrived in the parking lot bleeding all over the place.

> "Poirot," I said. "I have been thinking."
> "An admirable exercise my friend.
> Continue it."
> —Agatha Christie *(Peril at End House)*[xiv]

Pneumomediastinum: air trapped in the membrane that divides the thorax (the mediastinum).

So… **What're you gonna do about it?**

Choose from the treatment options.

a) Antibiotics

b) Bronchodilator

> A drug to dilate the lower airways in the chest.

c) Chest tube placement

d) Debride older wounds

e) Diuretic medications

f) Fluids

g) Mashed potatoes

h) Mediastinal transplant

> This is not really a thing (hint: don't pick it).

i) Oxygen

j) Pain medications

k) Probiotics

l) Splinting the ribs

> This is a way to stabilize multiple fractured ribs.

m) Staple his wounds

> Yes, we staple some wounds instead of suturing them.

n) Surgery to suture the wounds

o) Surgical exploration of the trachea

p) Surgical placement of a wound drain

> This is a device meant to help fluid from inflammation and infection drain from a wound while it heals.

q) Thoracic coupage

> Thwacking the chest to break up gunk inside.

STOP!!!!!!

Pick several treatments.

Which would make the most sense considering Asesino's wounds?

I would choose **a, d, f, j, n, o**! Possibly **p** as well (depends on what the wounds look like surgically).

Surgery to suture the wounds. You probably picked some surgical treatments, and that's certainly the right choice. Smaller, shallower wounds might heal well with just a few staples, but these big, deep ones are going to require some significant surgery to fix. It's very possible that you'll end up placing a drain in some of these wounds as well, which is a tube to help prevent fluid buildup in the initial phases of healing.

Surgical exploration of the trachea. In addition, you decide to do surgical exploration to explore the neck wounds and look for a tracheal tear. Many cases of pneumomediastinum will heal on their own medically (without surgical repairs), so this is unnecessary in every single case of pneumomediastinum. That may be because some cases involve tiny tears that heal on their own, or because the air is coming from somewhere other than the trachea (like a tear in the esophagus, extension from skin wounds, or microscopic lung damage). But since you're going to have your patient under anesthesia anyway (and don't like the idea of seeing this aggressive beast back again later), you'll go ahead and look for a tracheal puncture. A chest tube would be helpful if there were pneumothorax (air between the lungs and the body wall), but not for pneumomediastinum.

Debride older wounds. You'll also clean out and debride (cut away the dead tissue) the older wounds that look infected.

Fluids. IV fluids will help keep this guy's blood pressure up during anesthesia for surgery, which is even more important knowing that he's already anemic from blood loss.

Antibiotics and pain medication. You'll want to put this critter on some good antibiotics and stiff pain meds as well.

Mediastinal transplant is not really a thing, so sorry if you picked that option. And mashed potatoes *are* delicious, and are definitely a good idea, but they won't help much here.

You explain all of this (except the bit about the potatoes) to Arturo and ask him where all the older wounds on the dog came from.

"Oh…um…yeah… We live in the country, and there are coyotes and wild dogs around a lot. Asesino, he's my guard dog. He's seen a battle or two, you know? That's why I care so much about him! Do whatever you've got to do to fix him up, *gabacho.*[7]"

"I'll certainly do all I can, but it will not be cheap. I know you've got your son to think about too… There are ways we can cut back if needed."

[7] A somewhat derogatory word meaning "foreigner" sometimes used to describe someone outside one's own culture.

Arturo told you last time he was in that his 5-year-old son suffered from a rare neurologic disease and is on experimental medication. It's very expensive, as are the trips to the hospital to administer it.

The sparkle of a tear is visible in the tough-looking gangster's eye, as he mutters, "Appreciate that. Things are tight. But the medicine seems to help." He wipes away the tear, stares at the wall, and sniffles a little. "But don't worry about that. Got some cash from Dad's restaurant on the way here. Asesino's my best dog. Do what you need to do."

He wasn't kidding. Arturo pays his entire bill in wads of cash, then leaves. You stare morosely at the door for a moment, fleeting visions of mashed potatoes fading with your vanished lunch break, then turn to get to work.

Fred and Bojka set up a catheter, start some IV fluids, and give the first injection of pain medications and antibiotics. Then you scrub up, slip on your gloves, and step over to work your magic. It takes a while to place a Penrose drain (a type of latex tube that sticks out of a hole underneath a large wound to help it drain while it heals) and close the big wound in several layers. It takes more time to close the many smaller wounds, then debride the grody dead tissue from the infected wounds and close them up.

At last, you have your nurses turn over the patient to explore a medium-sized wound in the neck. You put some saline in and can see a faint hint of bubbling when he breathes. Time to look for a tracheal tear!

You've already spent a little time looking at relevant anatomy in a surgical textbook, so you expect it to look like this...

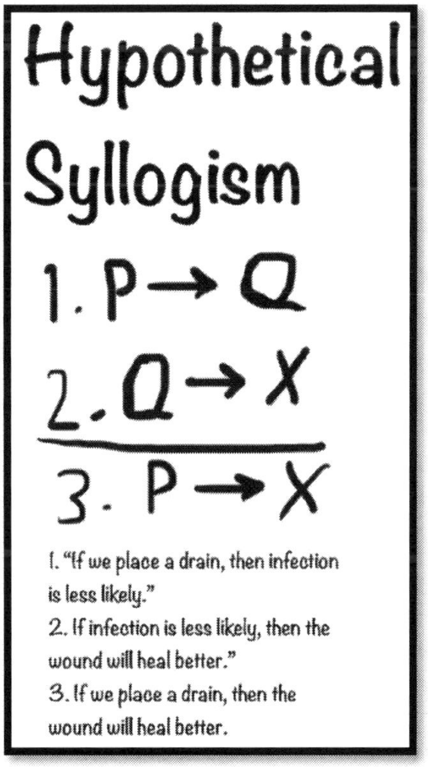

Here's a trachea as you see it in your surgical textbook...

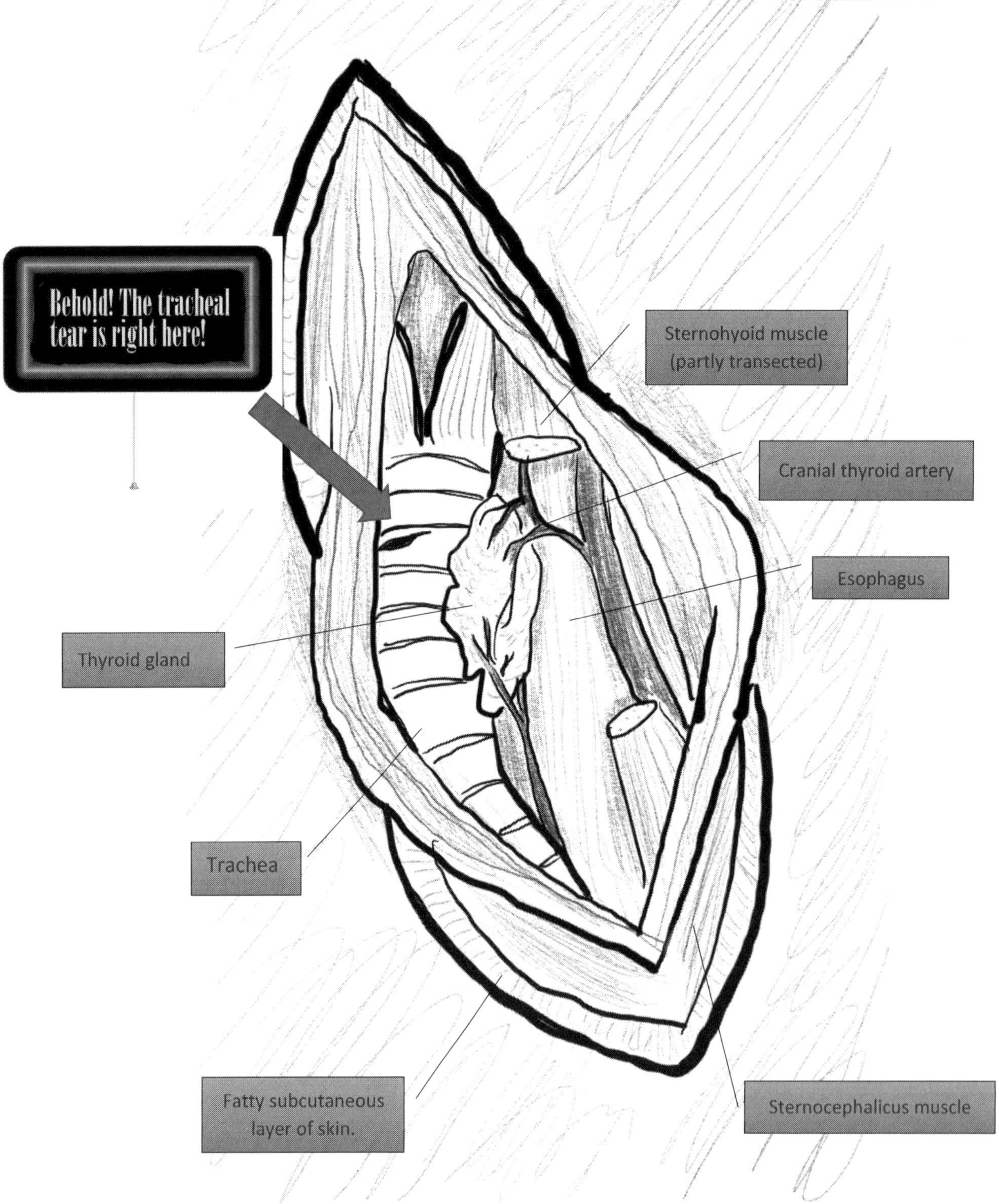

This being the annoying real world and all, here's what you actually see...

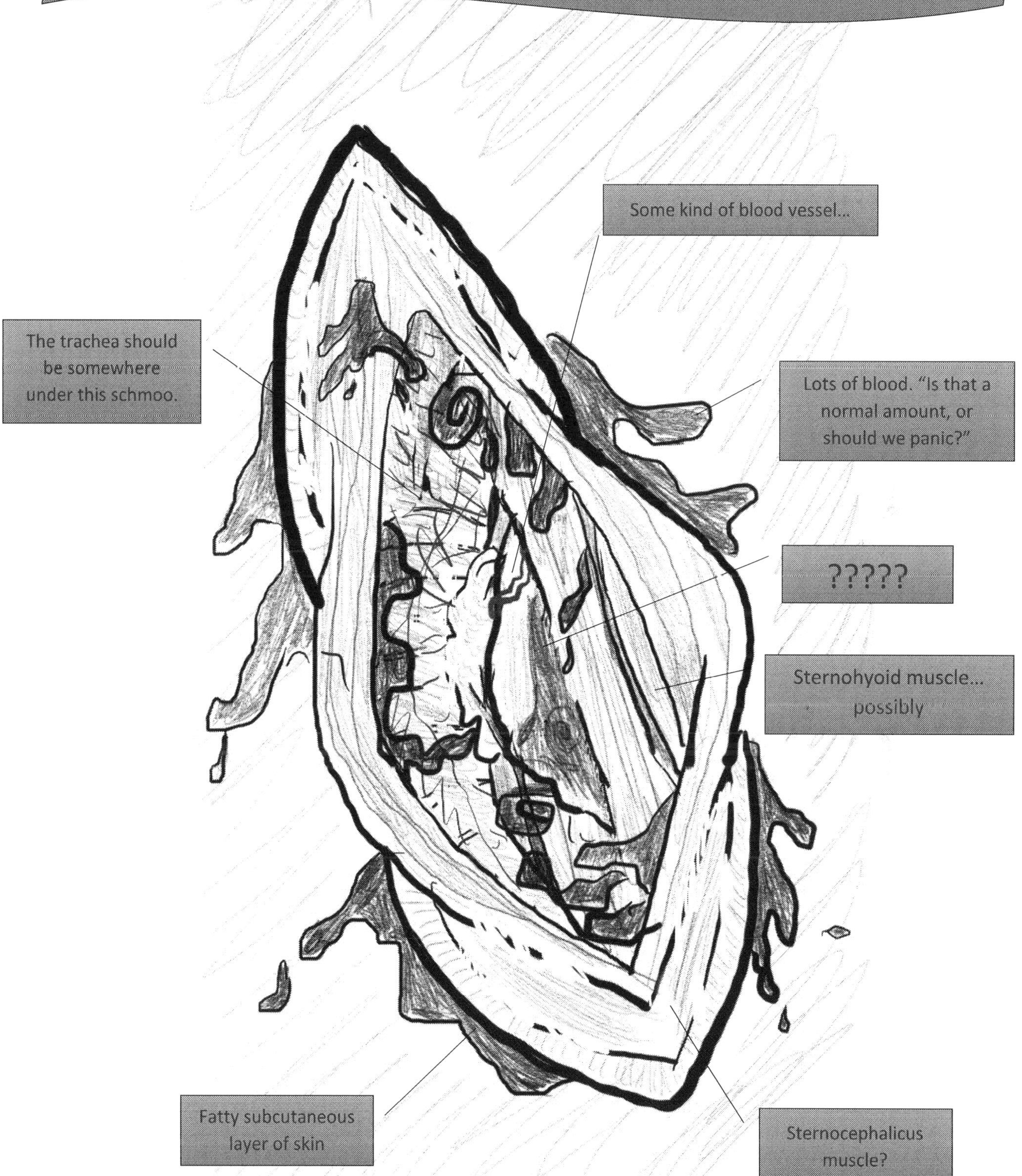

After a bit of sweating, muttering, and sloshing around in perplexing tissues and pools of blood (hoping you don't break anything important and praying you can put this mess back together again) you manage to find and fix a teensy tear in the trachea. Well done! You wake up Asesino, and he shows you his deep gratitude for fixing his windpipe by immediately trying to rip out your windpipe. Some good deeds never go unpunished.

But Arturo is very pleased with your work. And grateful. He hands you a bag from his father's restaurant, La Chuleta. "I felt bad for taking your lunch hour, amigo, and brought you some food from my father's restaurant. La Chuleta has the best smothered green chile enchiladas around!" The aroma of cheese and green chile is delightful, and you drool at that just like Asesino does at the smell of man flesh. You remember Mrs. Garcia from that morning and realize the restaurant is one and the same. Her comment earlier that morning (about how La Chuleta isn't seeing many customers) seems hard to believe. You thank Arturo with genuine feeling.

"You saved my best dog. Wish I could bring you more. But I gotta save everything I can for my boy."

You connect some more mental dots by this time and realize the grinning little guy Mrs. Garcia brought with her this morning and Arturo's sick son is also one and the same. "Hey, how is Samuel doing?" You didn't mean for the words to come out sounding choked.

A tear glistens again in those hard gangster eyes. "Not so good, doc, not so good. But my wife and I, we have hope that this new medicine he's on will make a difference." He nods his head and stares out the window at the parking lot for a minute. "I'll tell you something. This tough-guy thing I do…all this…it's all for little Arturo Jr. You know, to show him his daddy's strong, and is gonna stay strong for him. A man's gotta stay strong for his family, you know. He's gotta be willing to do anything for them. Anything!"

You suddenly feel ashamed for being so stressed about little things like missing your lunch and nod in sage silence (mainly because your mouth is full of delicious enchiladas). After you swallow, you offer to transfer Asesino to an ER for medical monitoring during the night while he recovers. Unfortunately, Arturo declines and decides to come back at closing time to take him home.

You take the picture Samuel drew for you of Gustav Mauler and tape it to a wall in your office. The thin, wobbly lines feel weightier now. You frown at the thought of the seriously ill little boy who drew them.

Then you guzzle some water to wash away that fabulous lunch before more clients come in.

Ah! And here they come, a-marching in one by one! Four booster vaccines, a puppy exam, and an angry cat with a malady.

Oh, joy!

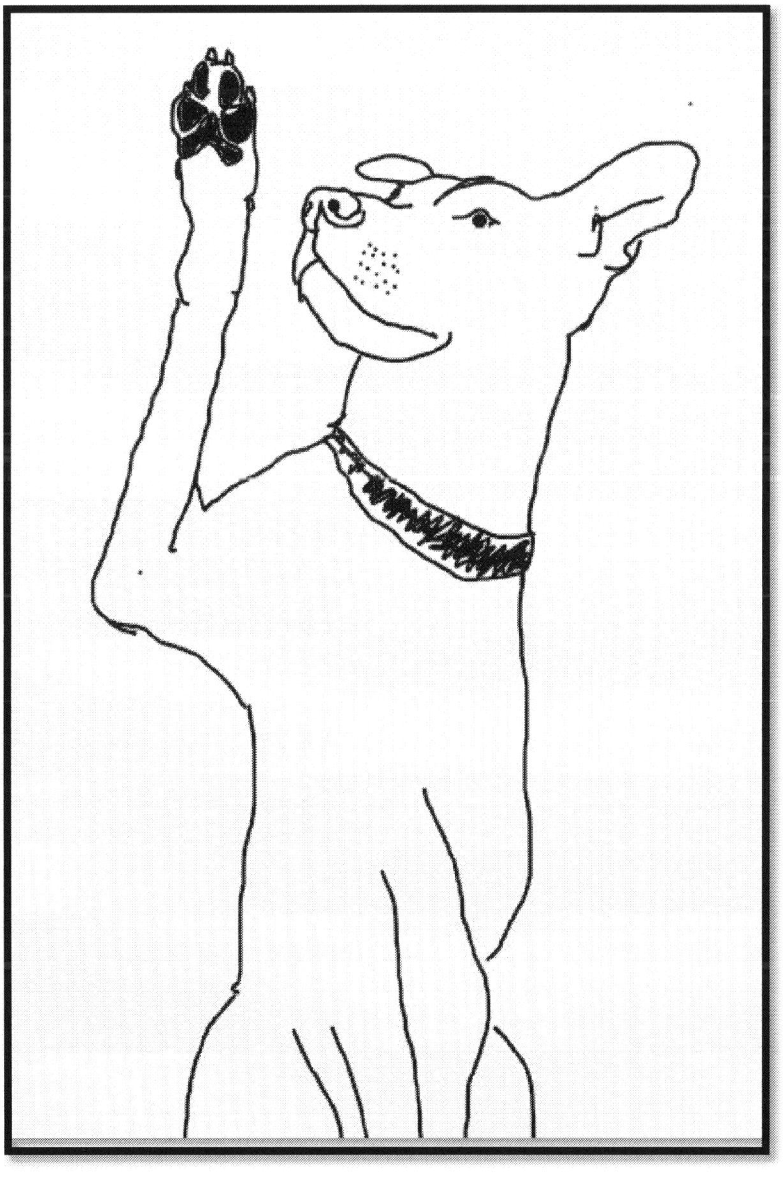

CASE #5

Your next case is a 10-pound cat who thinks she's a 150-pound mountain lion. Pearl is an old kitty in for acting lethargic. You don't notice a single shred of lethargy in the ball of fuzz and fangs that hisses and growls and rushes around the carrier the moment you get close.

"Haha!" the girl with Pearl laughs as you leap backward from the bundle of claws swiping at you from the bars. This girl is a thin, wispy specimen who looks unnaturally pale, as if she'd lived her whole life in a dark closet. But perhaps the black leather vest she's wearing over a frilled black shirt, with black pants and black boots and a black armband (along with black lipstick and black eyeliner), just makes her look pale by contrast. "Pearl doesn't like you much."

"Seems that way," you mutter. "Tell me what's going on with her."

The unidentified girl (who still hasn't introduced herself), flips a mane of purple hair over her shoulder and goes back to poking at her cell phone while she talks. "She's just not herself lately. Won't eat much. Doesn't move around much. Like she's the CEO of lazy."

"Is she having any vomiting or diarrhea?"

The girl holds up her slender finger to signal a pause for finishing a text, then answers, "Bet. She vomits every day."

"Ah. How about any coughing or sneezing?"

"No."

"Well…" you hesitate as you eye the snarling, spitting old demon in the carrier. "We're going to need to sedate Pearl to do an exam. We'll probably need to do some bloodwork too…"

She holds up that slender finger again to stop you. This time she even looks up from her phone. "Can I get one of those price sheet dealies?"

You blink a moment. "An estimate?" She gives a thumbs-up and an ironic smile. You go off to make her an estimate.

"What on earth are you eating?" you ask Bojka as you work at your computer. Your technician looks surprised and a little embarrassed.

"It's Cevapi. Balkan sausage. My Baba makes it. Want some?"

It looks a bit like a skinned squirrel, but you accept the offer. And you're glad you did: it's unexpectedly delicious. This day is turning from a culinary dirge into an absolute flavor exploration!

"Can you go over this estimate for Pearl?"

Bojka nods. A few minutes later, she drops a paper on the desk with a spidery signature on the "declined" line. "She has $7 to spend after paying for the exam, Doc."

Oh boy…

You go back to the room and explain there's no way you're going to be able to do an exam on this kitty without some serious drugs inside her. Her owner counter-explains she that cannot afford anything.

"Maybe I could, like, hold her for you? She's not that bad for me."

You have visions of your pale client covered in blood in the hospital calling her lawyer and wisely decline that plan. "Maybe that guy from the bookstore will pay my bill. He's cute. Can I get his number?"

"Um… No." And yuck. Bernard's old enough to be this girl's dad. You feel weirdly old yourself now, as you think, *What's wrong with kids these days?*

What are you going to do now?

Here are some options that go through your head…

a) Tell her to go away and take her devil cat with her
b) Skip the exam and send home some oral drugs and hope for the best
c) Take the carrier in the back and try to do an exam and some basic treatments with fast technicians and sturdy cat-restraining devices
d) Have a rage-fit and jump around and yell a lot
e) Do the sedation anyway with drugs that aren't technically yours and risk getting fired
f) Talk to your boss

Decision Making

In medicine, you have to make important decisions in real-world settings. Learn how to make logical decisions, even in cases when you have limited information (or those when you feel paralyzed by too much information). Here's a general 7-step process for making decisions…[xv]

1) Identify the decision to make
2) Gather information
3) Consider alternatives
4) Consider the consequences (risks vs. benefits)
5) Choose an option
6) Act
7) Review your action (maybe you can learn a better way for next time)

This is, frankly, one of the hardest parts of being a vet: having to make unexpected, important decisions in situations where there isn't really a good option available. You like your job, so you decide against option e. You like your fingers, so you decide against option c. You like feeling you're a nice guy, so you decide against option a. You like your career (which you'd put in jeopardy by treating with no exam), so you decide against option b. Option d is too juvenile, so you're left with **option f, going to talk to your boss**.

The boss is never happy about these things and fumes for a while about how hard it is to keep the clinic running when you want to give away free services. But eventually, you get the green light to do some free dexmedetomidine (a sedative), torb (for more sedation and pain relief), and a short blood panel, along with some simple outpatient treatments. You go back and explain to your client this glorious victory of beneficence you've won on her behalf.

She looks up from her phone for a second. "Dope."

You sort of expected a little more gratitude. But whatever, bruh.

Fred and Bojka look nervous as they don the thick, leather cat gloves and get ready to lift the carrier. You get some drugs ready in a syringe and crouch like you're about to hunt lions in the Savannah. One... Two... Go! They tip the carrier so that Pearl's scruffy flanks fall against the wall of it. You strike like an adder, jabbing her cat-butt with the needle. She explodes into a ball of teeth and terror.

But a few minutes later she's asleep, and you do your exam...

Step 1 – Organize the History

Signalment:

14 years old

Female spayed

Domestic long-haired cat

Presenting Complaints:

Lethargy

Vomiting

Horrible (that's a complaint from you, not the client)

Step 2 – Gather Clues

Physical Exam:

Abdominal palpation: Thin and "doughy," with no distension or masses noted

Attitude: Very asleep (thank the merciful Lord above!)

Cardiac: Slow heart rate (from sedation). Otherwise WNL (no murmurs or irregular beats).

Ears: Canals appear clear.

Eyes: WNL.

Integument: Decreased skin turgor (around 10% dehydration).

Lymph Nodes: Normal sizes.

Musculoskeletal: No significant findings.

Neurologic: No significant findings. But asleep… so how would you know?

Oronasal: Heavy calculus, otherwise WNL.

Pulmonary: Lungs sound clear. No increase in respiratory effort.

Urogenital: No significant findings.

Vitals: MM pink, tacky. CRT<2 seconds.

Not much to go on… **What do you think might be the problem?**

Differential Diagnoses:

Diabetes

A metabolic disease causing the body not to metabolize glucose normally.

Gastroenteritis

GI foreign body

Something indigestible sitting inside the intestines or stomach, which could cause a blockage.

GI parasitism

Hepatic failure

The liver failing to work (which could result from a number of different diseases).

Infection

Neoplasia (cancer)

Pancreatitis

Renal failure

Toxicity

Calculus (dental): a hard, porous coating over bacterial plaques on teeth. Synonym for tartar.

Decision time!

What tests do you want to run first?

Here's what you can do in this case…

- ✓ Short blood panel
- ✓ Short blood panel
- ✓ Short blood panel
- ✓ Short blood panel
- ✓ Short blood panel

What test do you want to run?

Hmm, how about a short blood panel?

"Use your eyes. Use your ears.
Use your brains—if you've got
any. And, if necessary—act."
—Agatha Christie[xvi]
(Death on the Nile)

Client: Suzzette Flandersby	Species: Feline	Sex: FS
Patient: Pearl	Breed: DLH	
Patient ID: 023577		

Chemistry 11 Panel

TEST	RESULT	FLAG	REFERENCE RANGE[xvii]	
ALT	30		8.3-35	IU/L
ALP	57		12-65	IU/L
Glucose	135	H	61-124	mg/dL
Total Protein	7.0		5.7-8.0	g/dL
Albumin	2.6		2.4-3.8	g/dL
Globulin	3		2.4-4.7	g/dL
BUN	102	H	15-31	mg/dL
Creatinine	7.1	H	0.5-1.9	mg/dL
Sodium	140	L	146-159	mEq/L
Potassium	3.1	L	3.8-5.3	mEq/L
Chloride	101	L	108-130	mEq/L

Right then, that's what you get…

In some cases, we must make educated guesses based on the limited information we have.

So, what do you think is going on with Pearl?

> If you feel a little stuck, be sure to check the chemistry panel explanation at the back of the book for more help with the bloodwork. Consider what explanation for the results would fit with an option on the differential list.

And **your diagnosis is?**

____

STOP!!!!!!

Yeah… You're probably right. This looks like some serious **renal (kidney) failure**.

An elevated BUN and creatinine together are called azotemia, and can have several causes. But renal failure is one of the biggest ones and fits the picture best in this cat. Older cats, especially females, are more likely to suffer from renal failure. It causes progressing nausea and vomiting. Renal failure can result in lethargy (as electrolytes are lost, and nausea mounts). Renal failure will lower all the electrolytes. And the other causes of renal failure are not usually gonna make the creatinine *that* high.

A couple of tests that could help confirm renal failure would include measuring the phosphorus, measuring something called SDMA, and doing a urinalysis (especially to look at the concentration of the urine). But of course, you can't run any more tests, so you'll have to work with what you have and roll with a *suspected* diagnosis (with informed reasons to suspect). You explain all of this to your client (while she twiddles with some mysterious social media app on her phone you don't recognize).

"K. So what now, dude?"

That's a good question, rude girl, that's a good question. Here are some possibilities. Keep in mind your treatments need to be cheap since your boss is paying for them and is irritated about it…

Universal Quantification

$$(x)(Hx \rightarrow Lx)$$

x = dog

H = has a heartbeat

L = is alive

An observation of a truth which is universal. A different way to state modus ponens type syllogism.

"If a dog (x) has a heartbeat (H) then the dog (x) is alive (L)."

Choose some from this list of treatment options.

a) Antibiotics

b) Blood transfusion

c) Deworming medication

d) Euthanasia

e) IV fluids

f) Kidney transplant

g) Magic kidney beans

h) Nausea drugs

i) Prescription food

j) Probiotics

k) Steroids

l) Subcutaneous fluids

MAKE YOUR CHOICE, ADVENTUROUS GUY!
ERE PEARL AWAKES TO SCRATCH YOUR EYE...

"As Shakespeare says, if you're going to do a thing you might as well pop right at it and get it over."

—P.G. Wodehouse, *Very Good, Jeeves!*[xviii]

In this case, the best things you can do are **h and l**. We can only treat the symptoms of failing kidneys (and can't reverse the underlying damage already done) and try to slow the progression of the damage.

Fluids. Treating dehydration and electrolyte changes with fluids is a good step. IV fluids would mean hospitalizing this murderous animal and are expensive, so you wisely do subcutaneous (SQ) fluids instead.

Treating nausea. You give a nausea injection and send home a few days of some oral nausea meds as well.

Kidney transplants in cats are not really a thing. And if it were, it would cost a lot more than $7. Kidney damage is irreversible and progressive over time in cases of chronic kidney disease in old cats. At least 75% of filtration function is compromised by the time you see any azotemia on bloodwork at all. And unfortunately, there are no magic kidney beans in stock at magicbeans.com.

Dietary changes can help slow the progression of the disease. You recommend a special diet with less protein, phosphorus, sodium, and more fiber. You also recommend an iron supplement, because (though you couldn't check for it today with your limited bloodwork) most cats with chronic renal failure are anemic.

"But…" you hesitate a little at this point. These conversations are always the tough ones. "Understand that Pearl's kidneys are failing, and that failure will progress to a point where she is sick and miserable all the time. It may well be at that point already." A group called the International Renal Interest Society (IRIS) makes guidelines for staging renal failure (deciding how advanced the disease is), often based on creatinine levels. A creatinine level over 5 puts Pearl in IRIS stage 4 for chronic renal failure: the most progressed stage there is. "She probably doesn't have long. And putting her down now would be a valid option to prevent her from suffering further." You remember something a client told you once: "Dogs and cats know they're hurting but don't know why. It's our job to decide for them when they've hurt enough."

She stares at the floor, and you wonder if that could be a tear in her eye. "No, bruh." She looks up finally, the hardness back. "She'll be fine. She's mean as $%@#. You sure I can't get Bernard's number?"

"No… Yes, I'm sure."

Well then, that's that. You do some outpatient treatments for Pearl, then wake her up.

Then Fred comes in to catch you up on your last appointment for the day.

Case #6

"We come at last to another day's end, and you're looking like a scruffy nerp-herder, Doc," Fred points to a streak of blood caked to your scrub tops, as you try to remember where you've heard that phrase before. Fred's a confirmed nerd and proud of it, so it's probably some movie with aliens or superheroes. "Dried blood will go about as smoothly with this next client as protons in a valence shell!" You blink in confusion, but Fred's enthusiastic grin always drags out an answering grin from your face, even when you don't get the joke. He hands you a fresh pair of scrubs he picked up from a pile in the back of the clinic. "Next client's a state senator, rich and snobby. Get cleaned up. Buck up too: you're almost done with the day!"

"Senator Dertoni brought in his 13-year-old Maltipoo named Harriet for coughing." Fred is walking down the hallway, reading his notes from the file as you head toward the bathroom. You can't brush the weird feeling that this feels like a political briefing on some drama show (except the ludicrous word "maltipoo" has probably never been part of any briefing). "Harriet's been coughing for the past three months on and off, but her coughing fits have been more frequent and sound nastier. She started breathing heavier than normal this morning, so he made an appointment. Her respiratory effort is not looking so great to me. I have Bojka watching her on some flow-by oxygen."

Senator Joe Dertoni is a big, broad-shouldered man somewhere in the tail end of his 50s. He's wearing khaki shorts and a polo shirt screaming tastefully but forcibly, "I'm a senator just off the golf course on my day off."

"Hello there! Thanks for meeting with me," he beams a big, political smile as you shake hands.

"Yes sir, absolutely," you reply, not exactly sure why. "I looked at Harriet on the way to the room and had Fred put her in the oxygen cage." Since she was in the treatment area with Bojka and sounded potentially critical, you did your physical before talking to the client this time.

Step 1 – Organize the History

Signalment:

13 yo

Female spayed

Maltese/poodle cross ("maltipoo")

Presenting Complaints:

Chronic cough

Dyspnea

> *"Political language is designed to make lies sound truthful and murder respectable, and to give an appearance of solidity to pure wind." –George Orwell.*[xix]

Step 2 – Gather Clues

Physical Exam:

Abdominal palpation: Suspect hepatomegaly.

Attitude: QAR (Quiet, Alert, Responsive).

Ears: Canals appear clean bilaterally.

Eyes: WNL.

Cardiac: Grade 4/6 murmur noted, with a PMI on the left.

Integument: No significant findings. Normal skin turgor.

Lymph Nodes: No significant findings.

Musculoskeletal: Crepitation and thickening noted in both stifles, the right elbow, and the left carpus.

Neurologic: No significant findings.

Oronasal: Severe calculus and dental disease.

Pulmonary: Lungs sound harsh in general, with crackles on the right side. Increased expiratory effort.

Urogenital: No significant findings.

Vitals: MM cyanotic, CRT<2 seconds.

You explain to Mr. Dertoni what you've found on the PE, using those mad medical term translating skills that have helped solidify your reputation for a good "bedside manner."

The senator nods sagely. "I can see the friend who recommended you was not being disingenuous." He smiles, and you thank him for the compliment, then hope it was really a compliment. Disingenuous is a word every politician is expected to say. But you're not exactly sure what it means. Maybe that's the point. "But do you really think she needs oxygen for a simple cough?"

Now you just feel frustrated that he plainly doesn't understand how serious a state Harriet is in.

Crepitation: a popping feel (in a joint's case, while you flex it). Can be a sign of arthritis in a joint.

Cyanotic: a lack of oxygen to the tissues, indicated by a purple color to the gums and tongue.

Dyspnea: abnormal breathing pattern.

Hepatomegaly: enlarged liver.

PMI: "Point of maximal intensity." The point where a heart murmur is the loudest.

QAR: quiet, alert, and responsive.

"I do think she needs oxygen. I'm more worried about her dyspnea than her cough, sir. She's really struggling to breathe."

"I see. Then let's cut to the chase: what's causing her symptoms?"

You ask him to tell you a bit more about her symptoms first, and he obliges. Harriet started coughing around three months ago. The cough progressed from an occasional hack to a moist-sounding wheeze over the last couple of weeks. She's been getting around less than usual, which her owner attributed to old age and arthritis. She's already on an NSAID pain medication and a joint supplement. She's not been around any other coughing dogs lately. No vomiting or diarrhea has been noted. She's been eating a little less than normal for the past week. She had one episode last week where she stumbled around for a few seconds and looked like she might fall down. But then she walked normally a minute later, so her owner didn't think much about it.

"Thanks," you say eventually. "Circling back around to your question, then, there are a number of different potential causes for her symptoms." (Wait… Are you sounding political now too? Is this catching?) "It could be anything from an infection in the throat, to a heart condition, to lung cancer. We should do some tests to help figure out what's causing her signs. She was oxygenating at 80% just now, so we should keep her in oxygen while we run our tests."

"I appreciate your candid language, Doctor. And to be candid myself, discretionary funds in my family are under the prevue of my spouse. So if you could generate a budgetary request for your proposed battery of diagnostics while I call her, I'd appreciate it."

"Oh…um…sure!" You leave the room and generate the proposed budgetary request. You send Fred in, armed with an estimate.

He comes back out looking just a little perplexed. "I think he said yes?"

He signed the approval part of the estimate, so it must be yes. Time to get some tests done! Running tests on critical patients needing oxygen can be risky. You have your techs check an sPO2 in oxygen and find it to read 93%. That means the supplemental oxygen from the O2 cage is helping the dog get better oxygen concentrations to her tissues. That also means she's currently "oxygen dependent," at least to a degree. It's possible to kill a patient trying to get radiographs, stressing them out while they can't breathe. So you instruct your nurses to take their time and get the diagnostics done in stages.

While they do, you work on your notes and ponder your differential list.

sPO2: "saturation of peripheral oxygen." A measure of how much oxygen is getting to tissues.

Differential Diagnoses:

Allergies

Bronchitis

An inflammation in the lower airways of the lungs.

Coccidiomycosis ("Valley Fever")

A fungal infection that attacks the lungs and bones.

Collapsing trachea

Weakness to the cartilage rings supporting a dog's trachea, which makes the trachea narrow on inspiration.

Congestive heart failure (valve disease, disease of the heart muscle, etc.)

Fluid in the lungs from the heart malfunctioning.

Heartworm disease

Neoplasia

Yersinia pestis

The bacterial infection that caused the Black Plague.

Pneumonia

An infection in the air sacs of the lungs themselves.

Tracheal infection

Tracheal irritation (from something irritating the throat)

Decision time!

What tests do you want to run first?

Here's what you can do in your clinic:

- ✓ Basic bloodwork (a chemistry panel and CBC)
- ✓ Blood gas analysis
- ✓ Clotting times
- ✓ Cytology
- ✓ Fecal floatation and direct saline exam
- ✓ FeLV/FIV test
- ✓ Giardia test
- ✓ Heartworm test (with rickettsial diseases)
- ✓ Leptospirosis test
- ✓ Pancreatic lipase (fPL or cPL)
- ✓ Parvo test
- ✓ Pregnancy test
- ✓ Radiographs (X-rays)
- ✓ Skin scraping
- ✓ Thyroid hormone level (T4)
- ✓ Ultrasound
- ✓ Urinalysis
- ✓ Wood's test or DMT plate

What tests do you want to run?

I say let's do some **bloodwork, a heartworm test**, and **radiographs**.

"The real world was far too real to leave neat little hints. It was full of too many things. It wasn't by eliminating the impossible that you got at the truth, however improbable; it was by the much harder process of eliminating the possibilities."

—Terry Pratchett, *Feet of Clay*[xx]

DA RADIOGRAPHS

Examine the radiographs on the next three pages…

The blank gray boxes highlight structures you might want to pay particular attention to.

Can you fill them in?

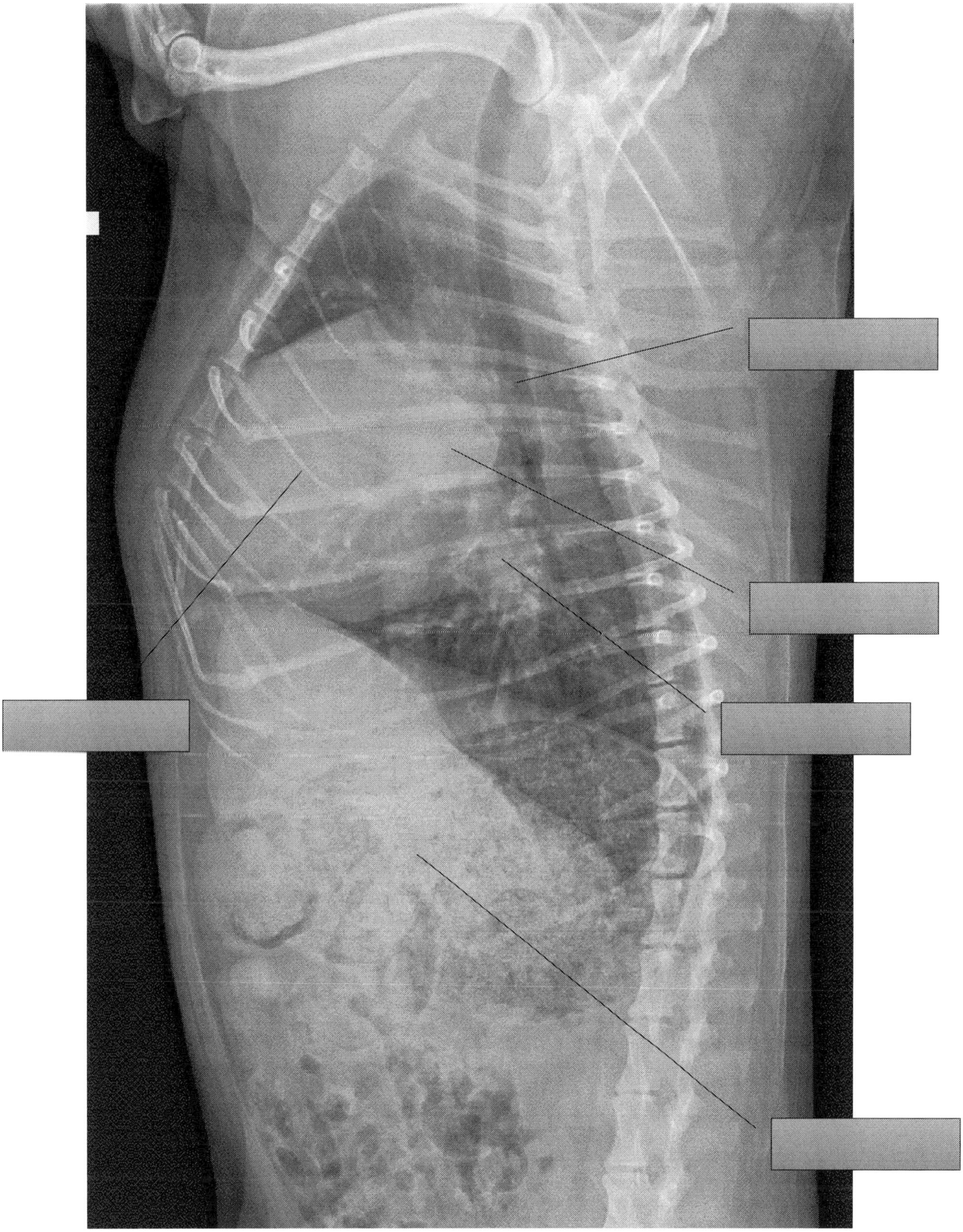

Left lateral view of the thorax and partial abdomen of "Harriet" Dertoni

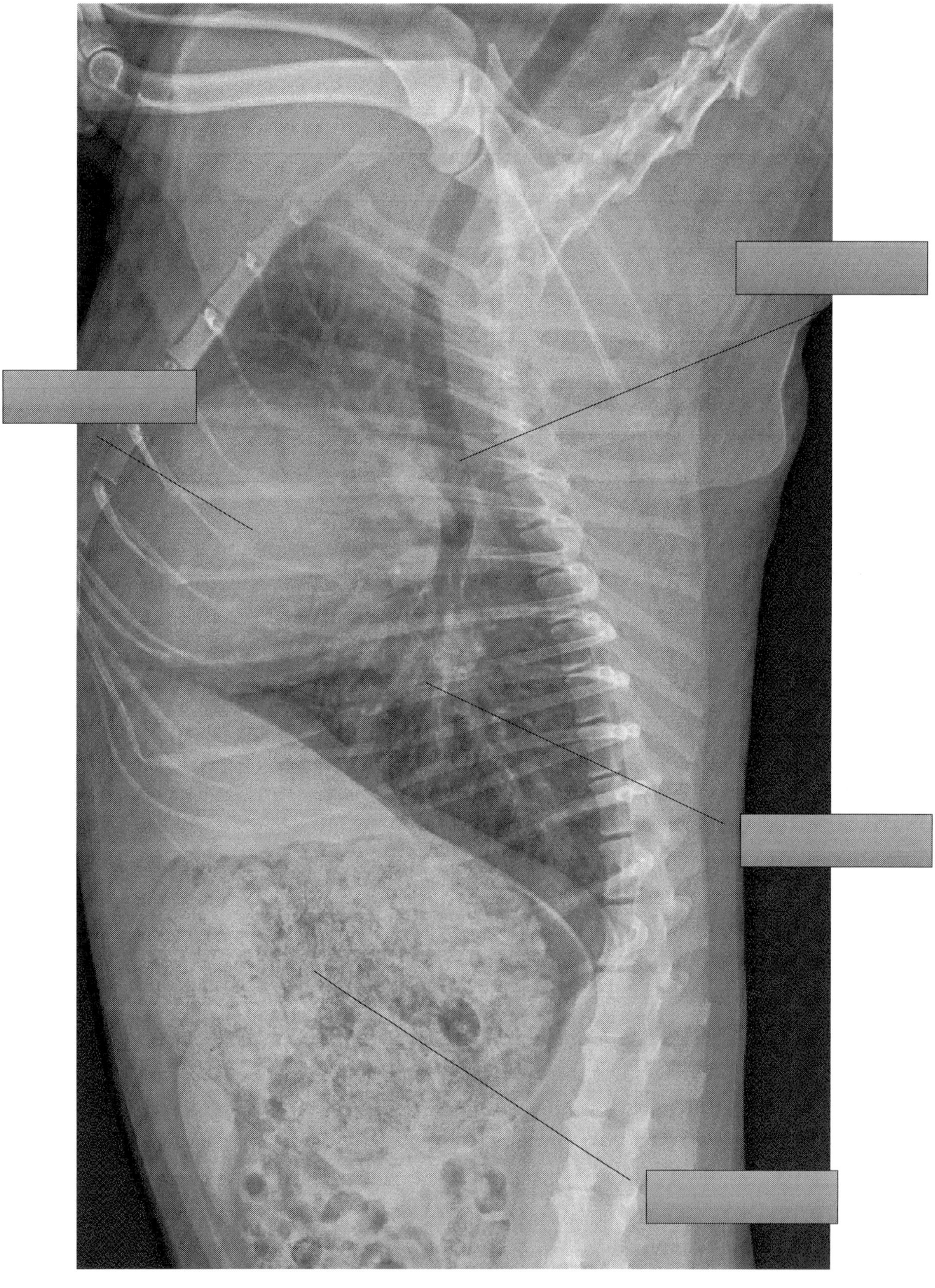

Right lateral view of the thorax and partial abdomen of "Harriet" Dertoni

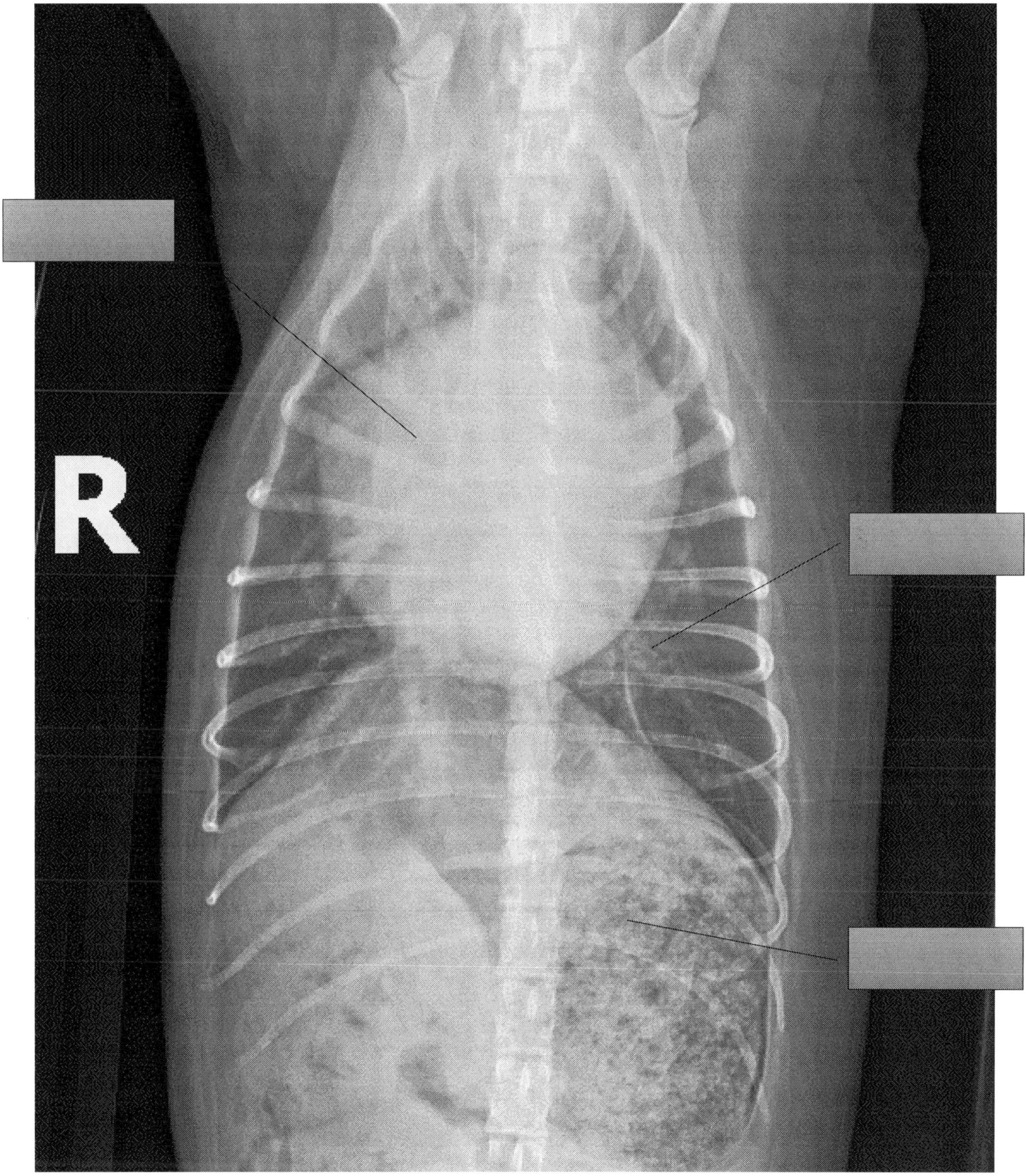

VD view of the thorax and partial abdomen of "Harriet" Dertoni

HERE'S WHAT STOOD OUT TO ME...

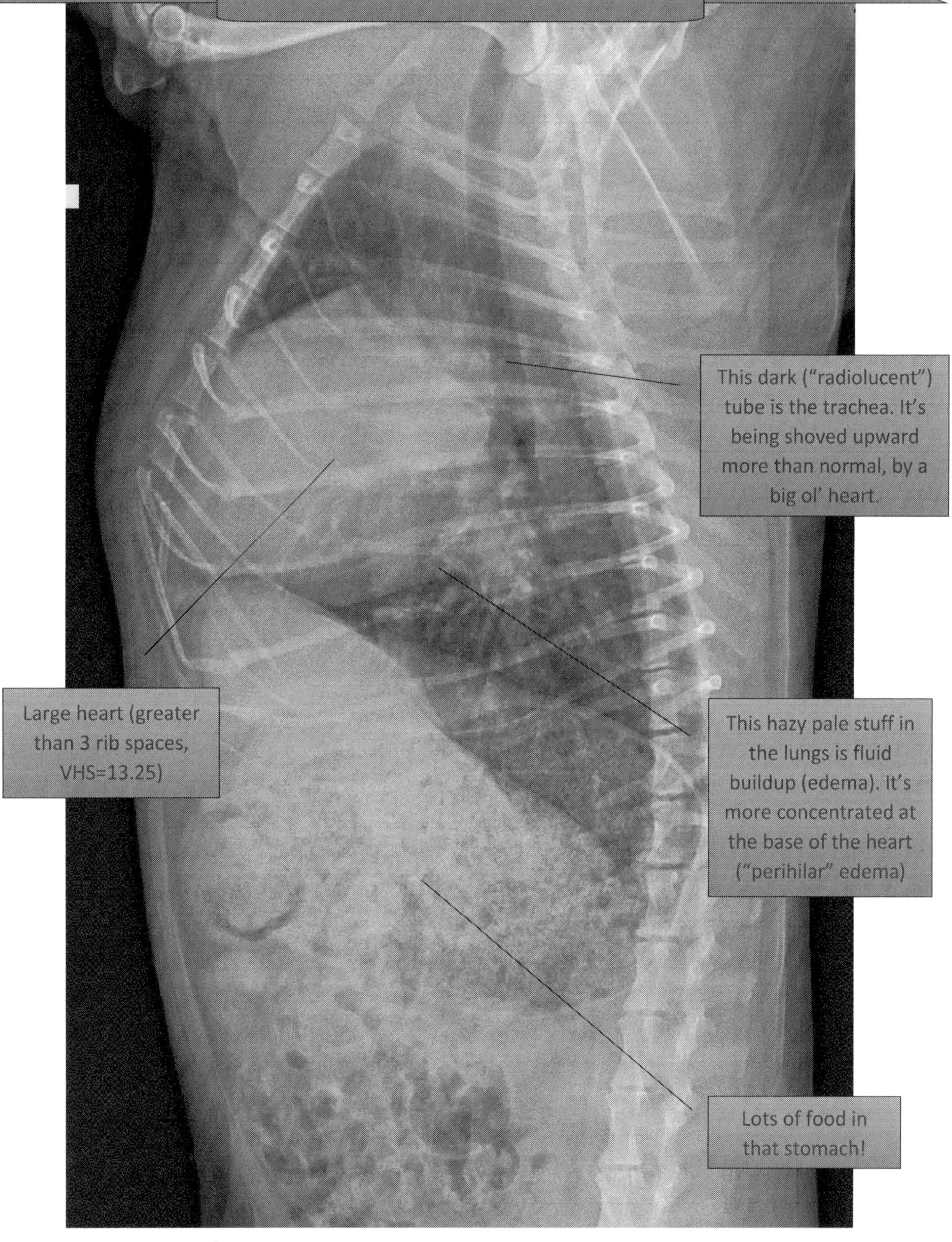

Left lateral view of the thorax and partial abdomen of "Harriet" Dertoni

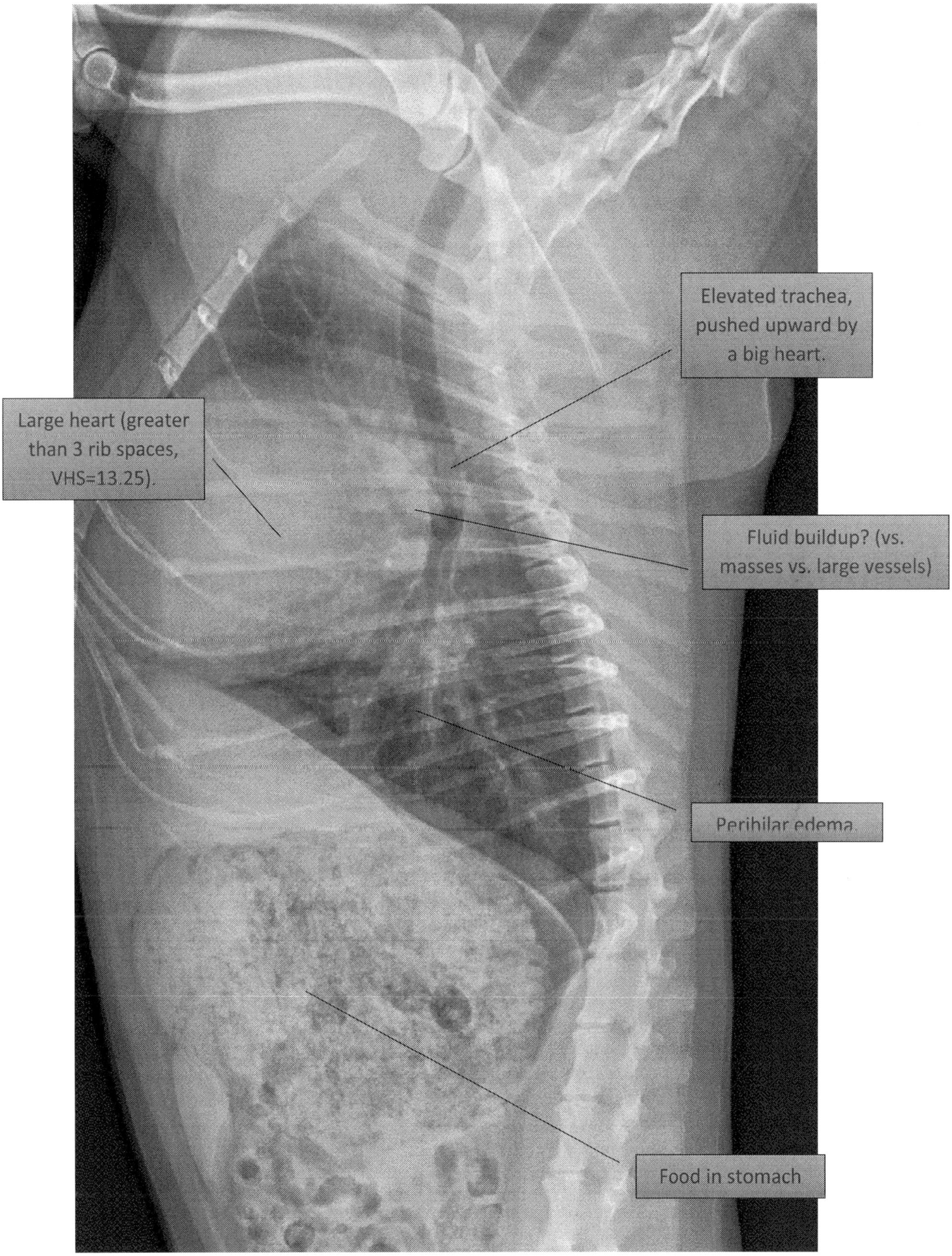

Right lateral view of the thorax and partial abdomen of "Harriet" Dertoni

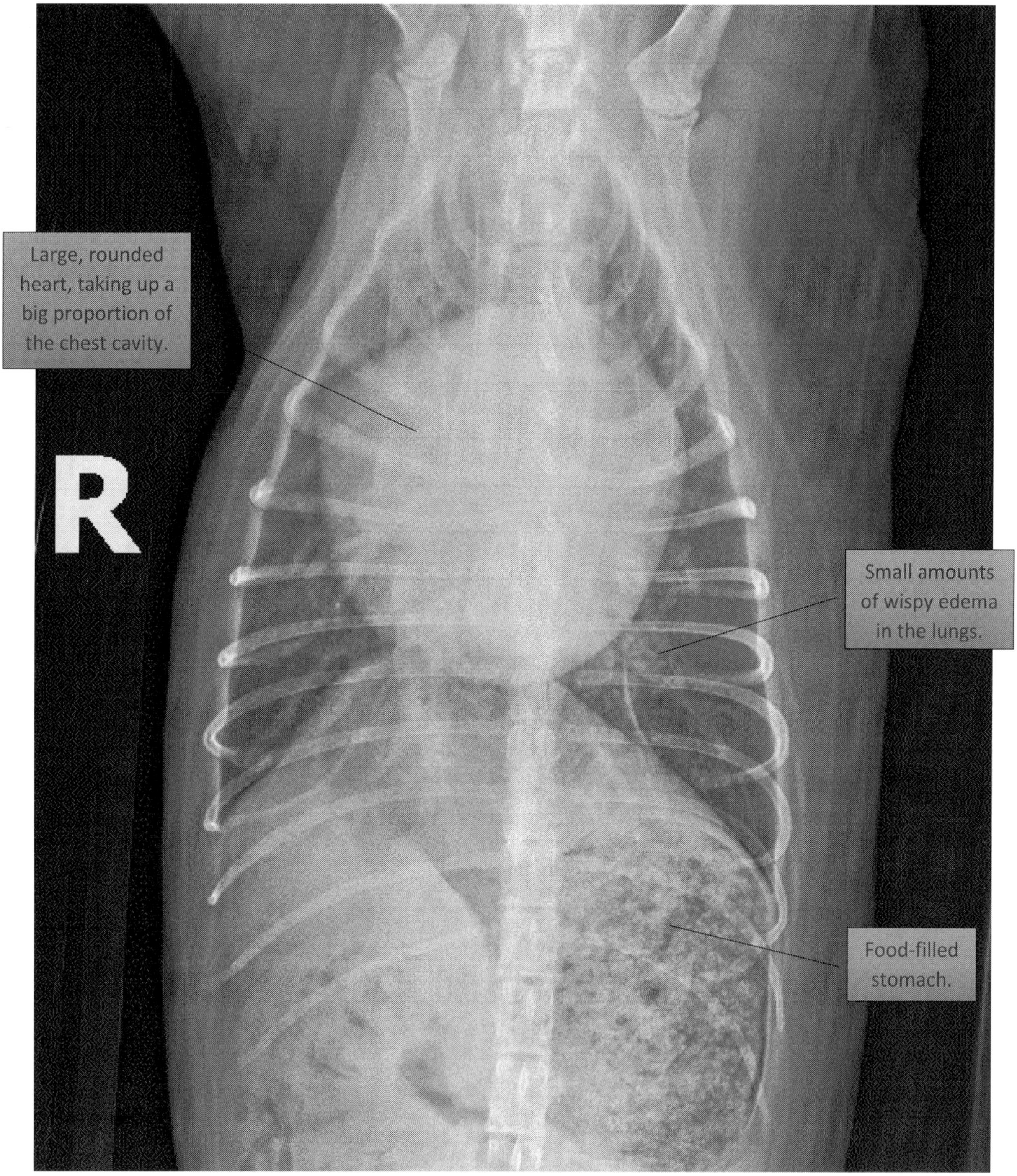

VD view of the thorax and partial abdomen of "Harriet" Dertoni

Da Bloods

Client: Dertoni, Vincent	Species: Canine	Sex: FS
Patient: Harriet	Breed: Maltese cross	
Patient ID: 026573		

CBC

TEST	RESULT	FLAG	REFERENCE RANGE[xxi]	
WBC	15		6.0-17.0	10x3/µL
Neutrophils	9.4		3-11.4	10x3/µL
Monocytes	1.2		0.15-1.35	10x3/µL
Lymphocytes	2		1-4.8	10x3/µL
Eosinophils	0.6		0.1-0.75	10x3/µL
Hematocrit	57	H	37-55	%
MCV	67		60-77	fL
Platelet Count	302		200-900	10x3/µL

Client: Dertoni, Vincent	Species: Canine	Sex: FS
Patient: Harriet	Breed: Maltese cross	
Patient ID: 026573		

Chemistry Panel

Test	Result	Flag	Reference Range[xxii]	
ALT	62	H	8.2-57	IU/L
ALP	150	H	10.6-101	IU/L
GGT	5		1.0-9.7	IU/L
tBili	0.8	H	0.1-0.6	mg/dL
Glucose	84		62-108	mg/dL
Cholesterol	119		116-254	mg/dL
Total Protein	7.1		5.5-7.5	g/dL
Albumin	2	L	2.6-4.0	g/dL
Globulin	2.5		2.1-3.7	g/dL
Calcium	9.9		8.7-11.8	mg/dL
BUN	67	H	8.8-26	mg/dL
Creatinine	2.6	H	0.5-1.6	mg/dL
Phosphorus	5		2.9-6.2	mg/dL
Sodium	129	L	140-154	mEq/L
Potassium	4		3.8-5.6	mEq/L
Chloride	102		102-117	mEq/L

And your diagnosis is?

You diagnose Miss Harriet today with **congestive heart failure**.

That's why her cardiac silhouette (her heart on x-rays) is bigger than it should be. It's filling up most of her chest, pushing her trachea upward, and spanning more ribs than it should.

What makes a heart get too big? The answer should be no surprise: there are lots of possibilities. There is, in fact, a differential list for the problem of an enlarged heart. But to simplify things, you tell Senator Dertoni that the problem (in old dogs especially) is usually a chronic degeneration of one or more of the heart valves. As those valves fail over time, fluid starts to back up because it's not flowing forward through the heart as it should. That causes fluid accumulation. Where the fluid accumulates depends on which side of the heart is failing. You see fluid buildup in the lungs (often at the perihilar area first) in left-sided heart failure (when the valve failing is on the left side of the heart). Fluid will back up to the abdomen and liver in right-sided heart failure (resulting in an enlarged liver and fluid in the abdomen).

Harriet's radiographs seem most consistent with left heart failure. There is fluid buildup in the lungs and no enlargement of the liver or free fluid in the abdomen (two signs you'd see with right heart failure). On Harriet's lateral view, the left side of the heart looks the biggest. You can get a rough idea of where to expect the four heart chambers, based around an imaginary clock face over the heart in lateral views.

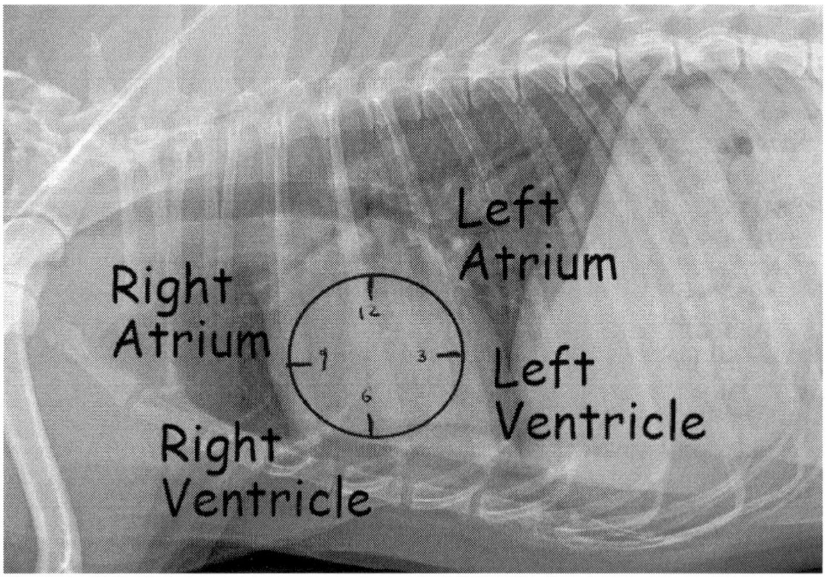

Harriet's VD view might indicate more right-sided heart or bilateral enlargement (both sides of the heart seem big). But it'll take an ultrasound of the heart (an echocardiogram) and maybe the abdomen to really be sure which part of the heart is failing. Ultrasound is an excellent way to look inside the heart and even characterize the lungs in more detail to help tailor treatment.

Cardiac Silhouette: the shape made by the heart on an X-ray.
Echocardiogram: ultrasound of the heart.

Unfortunately, when you point this out, Senator Dertoni raises the counterpoint that he doesn't want to pay for an ultrasound today. OK. You'll do your best with the inferences you can make from rads.

And the bloodwork? Screw on your thinking caps and imagine this…

As the fluid builds up in vessels, they start to get leaky, and fluid leaks out in all kinds of places. As fluid leaks out into tissues, there is a concentration of stuff normally floating around in that fluid within the blood vessels. BUN and creatinine go up, causing a "pre-renal azotemia" (azotemia not from kidney failure). That's why Harriet's BUN and Cr are high, but her phosphorus is normal.

Liver values can go up too, (that's why ALT, ALP, and bilirubin are high in Harriet), even more so when there's congestion of the liver (with right heart failure that Harriet probably doesn't have). Hemoconcentration is noted also as the fluids leak out and blood gets more concentrated (making Harriet's HCT high). Though sodium levels are increased overall (because the kidneys see more fluid they retain more sodium), it gets trapped in tissues and ends up lower than normal in the blood. That's why Harriet's sodium is low (though her Cl and K are not).

"We can't avoid reasoning; we can only avoid doing it well."

—Peter Kreeft, *Pocket Handbook of Christian Apologetics*[xxiii]

Hypervolemia: too much fluid (in blood vessels in particular).

Perilar: the area in the chest around the base of the heart.

Pre-renal azotemia: azotemia not from kidney failure.

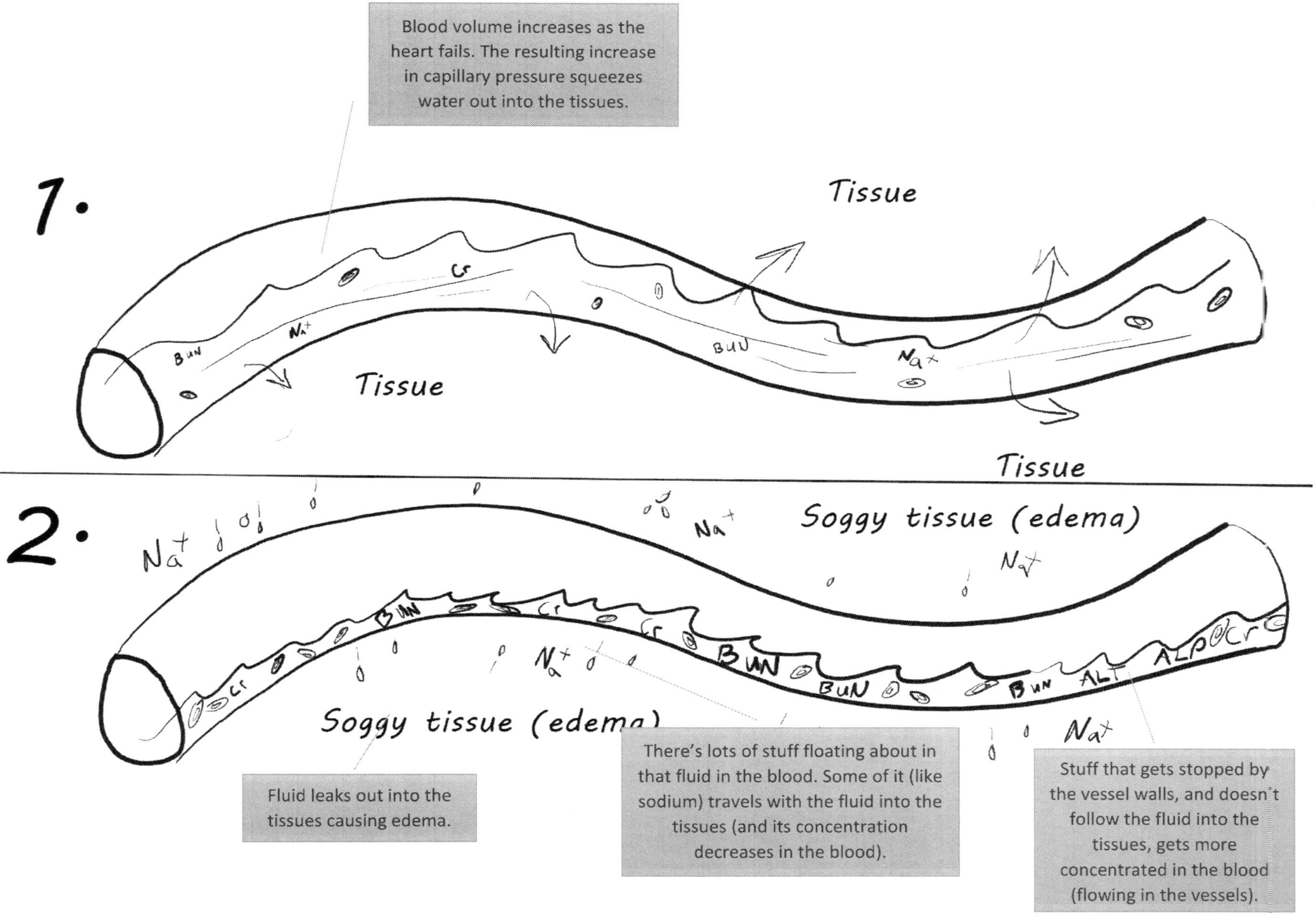
Blood volume increases as the heart fails. The resulting increase in capillary pressure squeezes water out into the tissues.
1.
Tissue
BUN
Na+
Cr
BUN
Na+
Tissue
Tissue
2.
Na+
Na+
Soggy tissue (edema)
Na+
BUN
Cr
Cr
Cr
BUN
BUN
BUN
ALT
ALP
Cr
Na+
Na+
Soggy tissue (edema)
Fluid leaks out into the tissues causing edema.
There's lots of stuff floating about in that fluid in the blood. Some of it (like sodium) travels with the fluid into the tissues (and its concentration decreases in the blood).
Stuff that gets stopped by the vessel walls, and doesn't follow the fluid into the tissues, gets more concentrated in the blood (flowing in the vessels).

If you were paying close attention to those gray boxes back at the radiographs, you may wonder what VHS means. It's not a reference to antique video players. VHS stands for vertebral heart score. This is a more objective way to measure heart size, by comparing the size of the cardiac silhouette to the length of thoracic vertebrae (the bones that make up the spine). You measure the heart's length from its stem to stern, then its starboard to port at its widest part (for all you yachters). Then you count the number of vertebrae that fit into those lengths together, starting from the edge of the 4th thoracic vertebrae (the thoracic vertebrae are the ones with the ribs coming off them, and big spines coming out from the top).

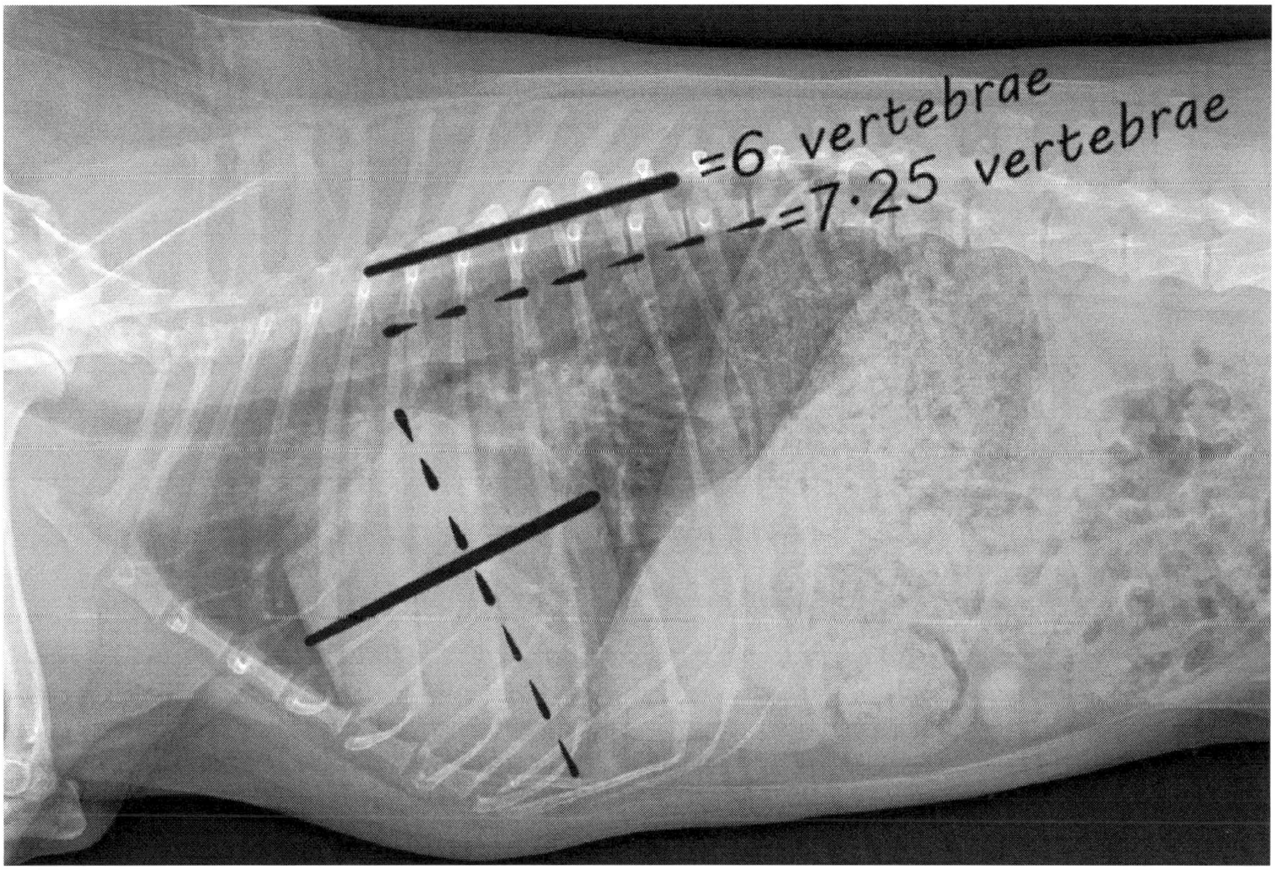

So, you've diagnosed Harriet with left-sided congestive heart failure.

The senator scratches his head at your brilliant lecture on the pathophysiology of heart failure. He looks confused. "Uh huh. I've heard your editorial, doctor but I motion to recommit. Let's lobby with less punditry and more layman's terms, shall we?"

Now you look confused. But only for a few minutes: you understand he's asking for a simplified explanation. "Harriet's heart is giving out 'cuz she's getting old. It's making fluid buildup in her lungs, which makes her cough and breathe heavily."

The light of understanding dawns upon his face. "Ah. How much longer does she have?"

You tell him there's no way to be sure: it may only be a few months.

The senator frowns. "Mrs. Dertoli won't like that. Any chance you would consider altering the diagnosis to something less…dire?"

"Um…no. Sorry sir, but me calling it something happier won't change the reality of her condition. And knowing what's really wrong with her helps us treat her signs."

He sighs and nods in a matter-of-fact manner that seems precisely calculated to impress news cameras that aren't there. "I do appreciate your candid and germane assessment of our situation, doctor. So, what remedy do we apply for?"

Good question, Senator Dertoni!

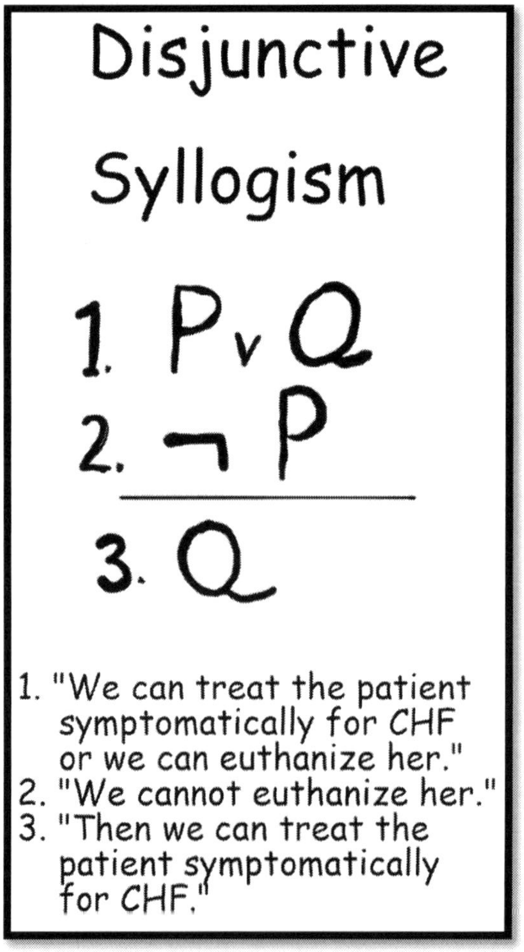

How are going to treat this beast?

Here are some choices....

a) An antibiotic
b) A diuretic injection
c) A diuretic pill
d) Heart transplant
e) Humane euthanasia
f) IV fluids
g) Medication to improve heart muscle contraction
h) Medication to slow the heart
i) Medication to speed up the heart
j) Milk thistle capsules for her liver
k) Nausea medication
l) A special diet
m) SQ fluids
n) A steroid injection
o) Time in the oxygen cage
p) A writ of *habeus corpus* to the General Accounting Office

"Unless you are good at guessing, it is not much use being a detective."
–Agatha Christie[xxiv]

You recommend **b, c, i, l, and o**.

Harriet needs **time in the oxygen cage**, with periodic **injections of a diuretic**, until her lungs are clear enough to get her breathing better and over this crisis. Then you'll send her home with **oral diuretics and medications to improve her heart's contraction**. She'll need a special diet to help her overall heart health. Nice job!

"Unfortunately, heart transplants aren't an option in dogs," you explain, "so we're left treating the signs of the failing heart and trying to slow the progression of disease with special diets and oral meds. But the progression of the disease is inevitable, and she'll eventually get so bad medications won't help. But that may be a couple of years into the future (in the best case)."

You also explain that, as nobody is in your clinic overnight, and you're getting ready to close, she'll need to spend the night at the animal emergency clinic after a few more minutes on oxygen and a furosemide injection (a diuretic).

The senator agrees and waits in the lobby while you and your techs get busy with Harriet. Not long later, Harriet seems to breathe just a bit better and is ready for transport. You go over a few last-minute directions while the senator nods and poses for any potentially hidden cameras.

"One more question, doctor… My daughter has a pet alligator named Joe, who she absolutely adores. He's been sneezing lately, and I wondered if you could take a look at him. Perhaps do a house call? I can remunerate you generously."

"Um… Senator, it's illegal to own alligators in New Mexico."

He slips an arm around your shoulder and lowers his voice. "Yes, and I appreciate you know your business so well. It's why you'd be the man for the job. You seem a discrete, non-partisan, head-on-your-shoulders kind of guy. Perhaps you'd consider helping poor Joe in exchange for a political favor."

"Um, no."

"You don't need any project grants, by chance?"

"No."

"Perhaps some lobbying work I could help with for a veterinary group?"

"No."

"Any referendums dear to your heart?"

"I don't think so?"

"Ah… I see." He takes his arm down and looks disappointed (and a bit worried). "How about we just pretend this conversation never happened then, eh? There's a mutual friend of ours involved, whom we wouldn't want to get into trouble…" You are perplexed and your face surely shows it (though you don't know it, and you *certainly* don't clap your hands).

But then Harriet arrives in Fred's arms, and you send her off.

The shadows in the parking lot are starting to look long, and you suddenly feel the weight of another busy day sitting its heavy end upon your weary head. You briefly dream about how you might have been home three hours ago if you'd only chosen a posh profession like a dentist or a chiropterologist. But your daddy always taught you to "take what you get, and don't throw a fit." So you sigh and drive home to watch your favorite observational comedy sitcom in a lonely living room until bedtime.

Riiiiiiing!!! BRRRRRRRIIIIIING!!!!

Your old nemesis, Alarm Clock, screams his nasty, blaring tones at you again with the crack of dawn. You jolt awake and stumble to the kitchen to jump start your heart with some coffee. Not many minutes later you pull into the parking lot of Noah's Haus Veterinary Clinic. There seem to be two police cars in the parking lot

The lobby is crowded with four police persons, who are questioning the technical staff and receptionist. Your boss walks up, and you read concern on his face.

"There was a break-in last night."

"What?"

"A break-in. When Amy got here this morning, she found the door forced open."

"Yipes. You OK, Amy?" Ms. Wallender nods and gifts you a sweet smile in exchange for your concern. "Was there anyone in the clinic?"

"No. The cash box was stolen, but all the drugs and equipment seem accounted for."

"Well, that's a relief." A bit of cash is, after all, a much smaller loss than a safe full of expensive controlled drugs. You need more coffee to deal with this and ask if you can go up to your office.

"Sure. The police may want to talk to you in a bit, but we'll let you know."

You journey up the stairs behind the reception counter, to where Mr. Coffee Pot lives on a little table on the landing. A few minutes later you sigh at your desk to take a sip. Then you notice something odd.

"Why's my computer on?" you mutter. You distinctly remember logging off and shutting it down last night after finishing up your records. In fact, the software Animal Haus uses for patient records is open as well. It's time to get your boss up here. The police are interested as well. They question everyone. Nobody at the clinic has opened your computer.

"It must have been last night's suspect," a muscular officer named Jerry Bylilly remarks. "Why would someone want to access medical records?"

You scratch your chin and sip your coffee epically. "Well, I suppose they might want information on a specific patient or client…"

"Or maybe they wanted to change something in their own record?" Bojka adds astutely.

"Yes. Could be. But they wouldn't be able to. Our software locks the records every night at midnight. When was the break-in?"

"Early this morning. We did pick up someone on the security camera breaking in at 3:20 am."

"Really? What did they look like?" Your interest is, as they say, piqued.

"Hard to say. They were wearing a ski mask and black clothes. Avoided the cameras like a pro (we only have a partial view of the one door camera he can't avoid). Average height and weight I'd say. Could be a girl or a guy."

That's not terribly helpful. And neither is the rest of the questioning. Eventually, it's time to see clients again, and the coppers wrap it up.

"We'll keep you informed on our progress," Officer Bylilly tells your boss, and you all thank them for their help and give each a donut before they file out the door.

Huh. Your mind ponders possibilities…

Who broke into your clinic?

What did they want?

Why did they take money from the cash drawer?

Why did they not take drugs?

Why did they get on your computer?

It's time to employ your expertly trained little gray cells…

CASE #7

Step 1 – Organize the History

Signalment:

3:20 am break-in

Forced door, professional avoidance of cameras

Average weight and height (male or female)

Presenting Complaints:

Broken door lock

Missing cash box

Accessed medical records on computer

A page of hints can be found in Appendix B...

Step 2 – Gather Clues

Physical Exam:

The front door: leads into the lobby. Made of some kind of light aluminum. A simple keyhole lock was picked.

Cash box: gone. Contained $200 in small bills (used for change), and copies of the day's receipts.

Millicent (your computer): In the same place she's always been, right there on your desk. Turned on. And your account is logged in to view medical records. The record on the screen was for "Mole Rat" Smiley. Looking through tabs, you find that all the clients from the past couple of days had been opened, as had the "settings" tab, the "preferences," your "account info," and the "permissions" tab.

Differential Diagnoses (for motive):

- ✓ Robbery (cash)
- ✓ Robbery (something they didn't find)
- ✓ Robbery (something they stole and we didn't notice yet)
- ✓ Coverup (trying to modify a record)
- ✓ Information (wanted a client's info)
- ✓ Information (wanted a patient's info)
- ✓ Information (wanted your info)

And your diagnosis is?

You don't really have answers at this point…

You'd like to think the intruder merely wanted some cash, but your open computer is troubling. Could they have stolen the cash hoping to distract the police from your computer? You can't think of any tests to run on this case (or, in other words, what to do next).

You'll have to content yourself with theories until something else turns up.

Something else does turn up, though it's not related to case #7. It's a fat lady with a trembling Yorkie.

She comes blustering into the clinic, clearly panicked, and practically throws her small, shaky creature at Fred. "Grizwalda's having some kind of seizure! Help her!"

Fred eyes you as he strides to the back with the quivering Yorkie. You follow him (while Ms. Wallender starts the lady's paperwork and Bojka gets some history).

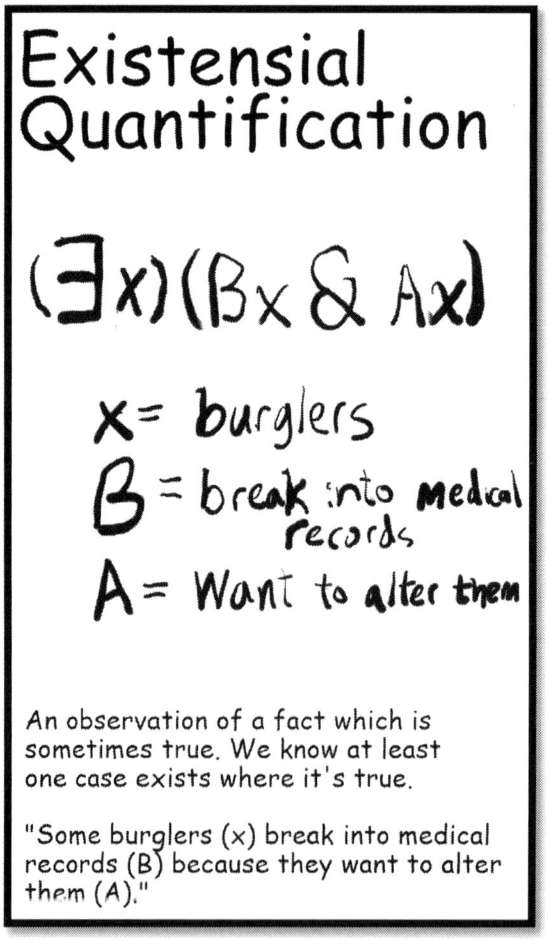

CASE #8

Sometimes the physical exam must come before a detailed history… You examine the quivering dog.

Step 2 – Gather Clues

Physical Exam:

Attitude: QAR.

Vitals: MM pale pink, CRT around 2 seconds. Body temperature 104.5°F.

Oronasal: A light brown nasal discharge noted from the left nares, moderate dental calculus.

Cardiac: Tachycardia (fast heart rate), otherwise heart sounds within normal limits.

Pulmonary: Lungs sound clear. Breathing fast, but no increased effort noted.

Abdominal palpation: WNL.

Ears: Canals appear clear.

Eyes: WNL.

Integument: Skin appears healthy, coat in good condition, normal skin turgor.

Lymph Nodes: WNL.

Musculoskeletal: Generalized, severe muscle tremors and stiffness. No pain on spinal palpation or neck flexion.

Neurologic: QAR, generalized tremors, all four limbs stiff and rigid but able to stand and walk. Pupillary light reflex (PLR) is WNL. Pupils are midrange in size and even, with no nystagmus.

Urogenital: Female, enlarged mammary glands bilaterally.

At this point, Bojka comes back with more history from the client (named Karen Midgewater). Grizwalda is a 4-year-old Yorkie who had a litter of puppies about three weeks ago. She's currently nursing all six of them. She acted a little off last week: coughing and sneezing, and her appetite was diminished, but she seemed fine otherwise. No vomiting or diarrhea has been noted. She's not currently on any medications. No exposure to any toxins has been witnessed.

What could be going on here? Time to make a differential list!

Differential Diagnoses:

- Central nervous system lesion — Something abnormal within the central nervous system (the brain and spinal cord). This could be any number of things, such as a stroke (interrupted blood supply) or a tumor.
- Electrolyte or mineral abnormalities
- Epilepsy — A condition characterized by periodic seizures.
- Infection (like meningitis, distemper, or rabies) — Meningitis is an infection in the brain or spinal cord.
- Liver failure
- Low blood sugar
- Toxicity (like mold toxins, drugs, tetanus toxin)

"With the Great Detective, to think was to act, and to act was to think. Frequently he could do both together."

-Stephen Leacock[xxv]

Mammary glands: breasts. M

Nystagmus: a continual twitc ... l, or rotary (round and round).

Pupillary light reflex (PLR): testing if the pupils constrict normally when a light is shone in the eyes.

Tachycardia: a rapid heart rate.

Decision time!

What tests do you want to run first?

Here's what you can do in your clinic:

- ✓ Basic bloodwork (a chemistry panel and CBC)
- ✓ Blood gas analysis
- ✓ Clotting times
- ✓ Cytology
- ✓ Fecal floatation and direct saline exam
- ✓ FeLV/FIV test
- ✓ Giardia test
- ✓ Heartworm test (with rickettsial diseases)
- ✓ Leptospirosis test
- ✓ Pancreatic lipase (fPL or cPL)
- ✓ Parvo test
- ✓ Pregnancy test
- ✓ Radiographs (X-rays)
- ✓ Skin scraping
- ✓ Thyroid hormone level (T4)
- ✓ Ultrasound
- ✓ Urinalysis
- ✓ Wood's test or DMT plate

What tests do you want to run?

Don't turn the page until you've picked some tests…

You decide to start with some **bloodwork**… An excellent choice!

There aren't many specific indicators to run the other tests on the list at this time. But remember, it's never wrong to pick another test as you get more information. In many cases with no clear answers, you'll find a clue on bloodwork (or other "broader" tests like survey radiographs), which you'll want to investigate further with different tests.

Da Bloods

Client: Midgewater, Karen	Species: Canine	Sex: F
Patient: Grizwalda	Breed: Yorkshire Terrier	
Patient ID: 026578		

CBC

TEST	RESULT	FLAG	REFERENCE RANGE[xxvi]	
WBC	19	H	6.0-17.0	10x3/µL
Neutrophils	14	H	3-11.4	10x3/µL
Monocytes	1.92	H	0.15-1.35	10x3/µL
Lymphocytes	3		1 4.8	10x3/µL
Eosinophils	0.6		0.1-0.75	10x3/µL
Hematocrit	56	H	37-55	%
MCV	67		60-77	fL
Platelet Count	302		200-900	10x3/µL

Client: Midgewater, Karen	Species: Canine	Sex: F
Patient: Grizwalda	Breed: Yorkshire Terrier	
Patient ID: 026578		

Chemistry Panel

TEST	RESULT	FLAG	REFERENCE RANGE[xxvii]
ALT	45		8.2-57 IU/L
ALP	56		10.6-101 IU/L
GGT	6.5		1.0-9.7 IU/L
tBili	0.5		0.1-0.6 mg/dL
Glucose	72		62-108 mg/dL
Cholesterol	221		116-254 mg/dL
Total Protein	6.2		5.5-7.5 g/dL
Albumin	3.1		2.6-4.0 g/dL
Globulin	4.5	H	2.1-3.7 g/dL
Calcium	4.2	L	8.7-11.8 mg/dL
BUN	24		8.8-26 mg/dL
Creatinine	1.1		0.5-1.6 mg/dL
Phosphorus	5		2.9-6.2 mg/dL
Sodium	145		140-154 mEq/L
Potassium	4.3		3.8-5.6 mEq/L
Chloride	104		102-117 mEq/L

Well, there you have it… Bloodwork results!

You have astutely noticed the most significant abnormality on this chemistry panel: Grizwalda has very low blood calcium (hypocalcemia). We measure calcium in two ways: total serum calcium or ionized calcium (our blood machine measures total serum calcium). Calcium below 6 mg/dL will produce clinical signs (things like muscle tremors, cramping, and restlessness). Calcium below 4.5 mg/dL will cause severe signs (like tetany, heart arrhythmias, lethargy, seizures, and death) and is considered life-threatening.[8] Grizwalda's total blood calcium 4.2 mg/dL!

What do you think for a diagnosis?

You'd better think fast because this little critter is shaking and fasciculating rougher'n a California earthquake. Her owner is pacing back and forth in the lobby and grumbling at the staff that she wants to go back and see her little Grizzy Baby.

Don't panic: but do have a plan. Time to take this bloodwork and explain things to Mrs. Midgewater before this situation gets worse…

And your diagnosis is?

STOP!!!!!!

Don't turn the page until you've made a diagnosis…

Hypocalcemia: low calcium levels.

Ionized: having a charge on it (positive or negative molecules or atoms). *Not* the act of changing one's name to Ian.

[8] Death is, in fact, almost always considered life-threatening.

What might have made Grizwalda's calcium get so low?

That's right! Nursing six puppies did it! That's why this type of hypocalcemia is often called *pregnancy eclampsia*. You probably remember your momma always saying, "drink yo' milk boy: it's chock fulla calcium for strong bones!"

Well, it turns out (in this instance) that she was absolutely right. Mammalian milk is full of calcium: calcium that producers of milk (in this case Grizwalda) are losing from their body. Normally a dog momma would make up for that by eating a lot of calcium. And normally the body uses parathyroid hormone to tell the intestines to absorb more calcium, and the kidneys to excrete less calcium.

But sometimes we owners don't offer a sufficient diet for that (because we don't realize what she was eating before the puppies came isn't adequate afterward), and calcium gets low. Sometimes hypocalcemia is triggered by other diseases, like parathyroid disorders (problems with the gland where the parathyroid hormone is made) or pretty much any disease that makes a dog eat less. What do you think triggered Grizwalda's hypocalcemic episode today? (Look for an answer in the appendix.)

Why would low calcium cause so many problems?

Calcium ions (a charged calcium) are needed to make muscles contract properly. The myosin and tropomyosin molecules that make up a muscle fiber are designed to interact with each other based on the flooding of calcium ions from a cell. If there's no calcium in the blood, the muscle fibers do all kinds of weird connecting and disconnecting, and abnormal twitching of muscles happens. This causes fasciculation and tremors, as well as stiffness and cramping all over the body. That cramping produces tons of energy (from ATP splitting, etc.), which makes heat, which makes a fever.

Eventually, all this together can make seizures happen. Remember from anatomy class that there are several important kinds of muscles too. There are voluntary muscles and involuntary muscles (those *really* important ones that control your heart for pumping blood and your diaphragm to help you breathe). When calcium gets low enough to affect those muscles, you get heart arrhythmias, progressing to a stopped heart and paralysis of respiratory muscles: and you die! And that's just bad news.

Calcium's effect on the body is an amazing example of physiology! Physiology is the study of the body's function (as opposed to its structure, which you study in anatomy). You get to spend countless hours puzzling over it at vet school. In all seriousness, you get to spend countless hours *wondering* about

Parathyroid hormone: a hormone (made in a gland called the parathyroid gland, which is in the neck near the thyroid gland), which regulates calcium.

physiology more than puzzling over it. The microscopic complexity built into each little furry creature that walks into your clinic is truly mind-boggling. And we still only know half of it!

One wonder of clinical medicine is that it's the point where physiology, pharmacology, anatomy, and entire lists of other "life science" disciplines get applied to real life. You, brave clinician, get to take all that wild, mind-boggling theory swirling about in your bewildered brain cap and use it for something good in the real world! Pretty cool!

Well… Mrs. Karen Midgewater is less impressed. She doesn't care about actin or myosin or calcium receptors or ATPase. She cares about Grizwalda and wants you to stop babbling and make her better before she dies. That's a good idea, but a perplexing conundrum.

How will you make this shaky little, calcium-deficient, febrile, dying Yorkie better?

Choose from this list of treatment options:

a) Antibiotics
b) Anti-seizure medications
c) Cooling measures
d) Dextrose — A form of blood sugar you can give IV for low glucose.
e) Diuretics
f) Get her a new set of muscles
g) Give her calcium
h) Hormone therapy
i) IV fluids
j) Mannitol — A drug with a big ol' molecule that pulls fluid out of the brain into the vascular space.
k) Steroids
l) SQ fluids
m) Sedatives
n) Wean the puppies

MAKE YOUR CHOICE, ADVENTUROUS BUM,

ERE GRIZZY BABY DIES FROM LOW CALCIUM…

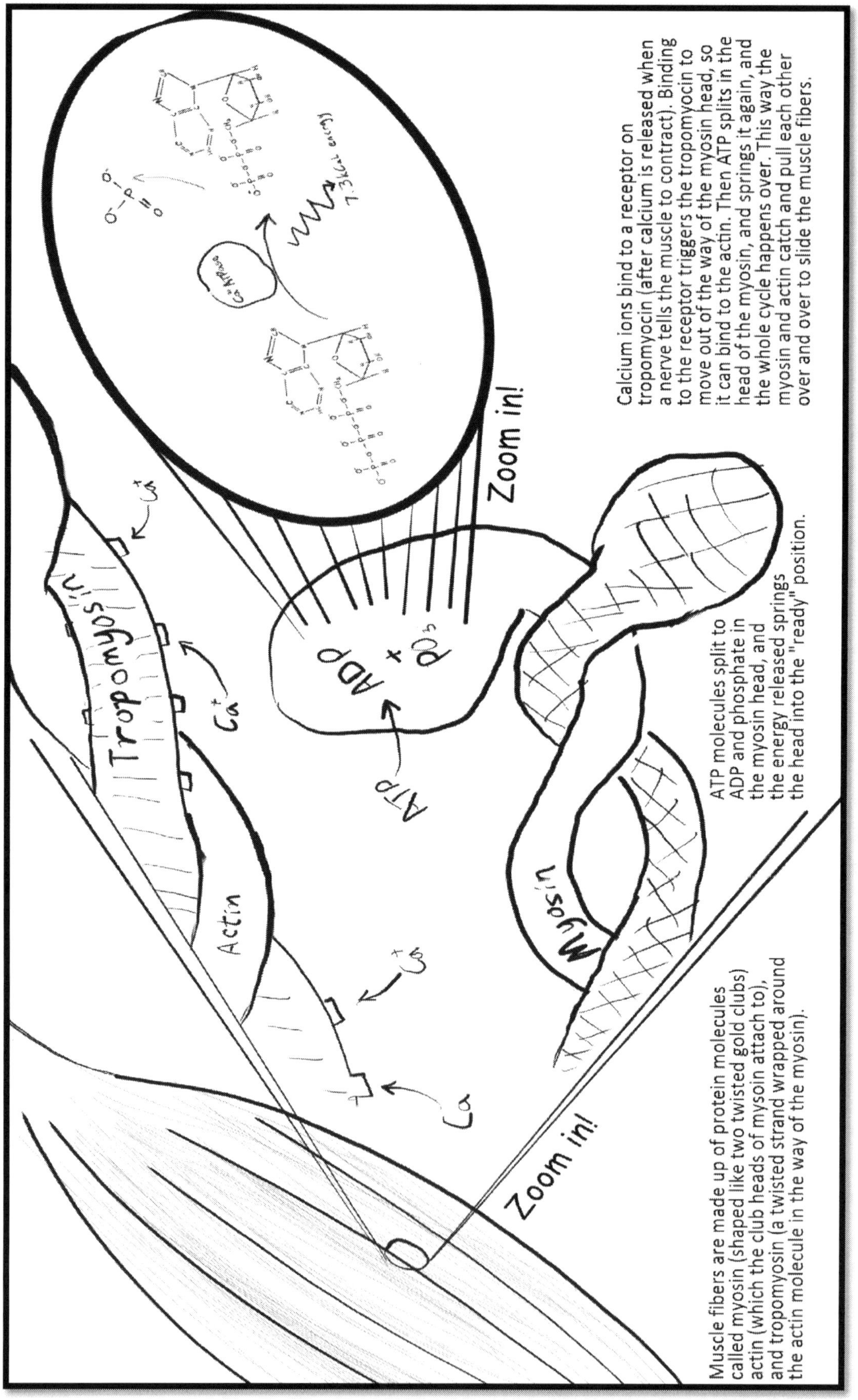

Calcium's effect on muscle fibers looks just a little like this—just a teensy sample of the wonders of physiology!

You astutely chose **c, g, and n.**

Give her calcium. This seems a logical treatment for low calcium levels! In the clinic, the form of calcium you use is calcium gluconate, and you have to give it carefully and slowly. Give too much of it too fast and you can stop her heart (something that's generally frowned upon in clinical settings). You give a dose of calcium gluconate in saline slowly IV while monitoring her heart with an ECG machine.

Cooling measures. These include a fan (which Fred already started), sometimes wetting her fur, and sometimes putting some ice packs under towels beneath her.

Anti-seizure medications are sometimes needed in these cases. But Grizwalda is still conscious and not to the point of having seizures. There are some severe cases where you need mannitol as well, for swelling in the brain, but we'll wait and see how Grizwalda responds to calcium gluconate first. Happily, Grizwalda responds well to your calcium gluconate. You can see a marked improvement within minutes after her dose. She's calming down, her muscles are trembling less, her heart is not racing anymore, she's not stiff and can walk better. You pick up the bottle of calcium gluconate and kiss it. "That's some gooooood stuff!"

Wean the puppies! Mrs. Midewater is happy too… Until you tell her she should bottle-feed those six puppies and wean them early. That requires work: lots and lots and lots and lots and lots and lots of work. She'll need to bottle-feed them a commercial puppy milk replacer (formula) every 2-3 hours around the clock. But it's certainly the best thing for Grizwalda. She should also supplement oral calcium carbonate (a common antiacid for human stomachs you get over the counter).

There's one thing on the list you shouldn't pick. If you picked steroids… For shame! For shame! Steroids are useful for many conditions, but not this one. They will make things worse by lowering calcium levels further (by telling the kidneys to excrete more of it, and the intestines to absorb less of it).

Well, good job! Soon Grizwalda's on her merry way out the door. Karen complains about her bill but pays it after a fight. She complains about your bedside manner. She complains about having to bottle feed the puppies. She's evidently a complainer. But you don't have time to stress about it because your next appointment is here already.[9] It's a 4 month old Labrador who needs three vaccines but won't hold still. It takes a few tries, and you nearly jab Fred in the toe with a needle after the lab's tail flicks it out of your hand. But you get the job done.

"I have a chubby puppy for you to see in room 3," Bojka tells you as you wave farewell to Mr. Hatahle and his bouncy Labrador pup…

[9] So, you file it away in your mental "to-stress-about-at-1-am-for-no-reason" drawer. A lot of your cases get filed there.

Case #9

She wasn't kidding. There's a seriously chubby puppy groaning on the table when you walk into the room. A chubby young lady is rubbing its fuzzy noggin vigorously. She introduces herself as Ember Daydream. She's wearing an excruciatingly bright, tie-died dress and peace-sign earrings. She's also wearing a pleasant smile, which you return.

"This is Ocean," she introduces the fluffy pooch. Ocean blinks at you, walks across the table, and stands up to put two paws on your arm. He wags his little tail in introduction, while you try to work out how many dog breeds he's built from.

"Hi there, Ocean! What seems to be the problem with you today?"

"I'll translate," Ember offers matter-of-factly. You nod, trying to look serious so you don't laugh and offend her. "I just feels icky in general today, like my aura's all off. And my tum-tum is all big." She puts her hands over the pup's ears and whispers, "He had some diarrhea too, and just seems a bit burpy." Then she scratches her head and waits for you to reply.

"Burpy? Have you vom…" You color just a bit as you realize you're talking directly to Ocean now. You turn your gaze pointedly back to address his owner. "Have you seen any vomiting?"

"No. But he's drooled a lot, and once he retched like he was going to vomit."

"Is he current on his vaccines?"

"No," she replied, scratching her right arm. "I believe in a holistic, natural approach to health, doctor, and am still trying to make my mind up about vaccinating him."

"Ah… Has he gotten into anything unusual lately? The trash? Chewed up any plants? Been around any pills?"

She picks Ocean up and puts his forehead against hers for a moment, then answers, "No."

"Alrighty then. Let's see what we can find on an exam then, shall we?"

Step 1 – Organize the History

Signalment:

3 month old

Mix breed dog

Male intact

Presenting Complaints:

Decreased appetite

Distended abdomen

Malaise — Feeling icky

Nausea

"He may be mad, but there's method in his madness. There nearly always is method in madness. It's what drives men mad, being methodical."
—G.K. Chesterton[xxviii]

Step 2 – Gather Clues

Physical Exam:

Abdominal palpation: Abdomen appears distended visually, and the patient is tense and painful on palpation (maybe more in the cranial abdomen than the caudal abdomen). No obvious mass or foreign body is appreciated (though you really can't tell because he's so tense).

Attitude: BAR (Bright, Alert, Responsive).

Cardiac: No murmur or arrhythmia noted, strong femoral pulses.

Ears: Clean canals bilaterally; WNL.

Eyes: Clear bilaterally; WNL.

Integument: Slightly decreased skin turgor noted; crusty dandruff noted in large amounts over his dorsum (back) with some thinning hair over his back as well; seems pruritic (itchy) in general; um…it could be your imagination, but it looked like some of that dandruff moved on its own.

Lymph Nodes: WNL.

Musculoskeletal: WNL.

Neurologic: WNL.

Oronasal: Clean, healthy teeth and gums, no nasal discharge, ptyalism (drooling excessively).

Pulmonary: Mild tachypnea, lungs sound clear, no increased respiratory effort.

Urogenital: WNL.

Vitals: MM pink and moist, CRT<2 seconds.

Dorsum: the back, along the spine.

Ptyalism: excessive drooling. You could also say "hypersalivation," but it's not nearly as fun.

Tachypnea: fast breathing rate.

Hmm… You're picking up on more than one problem after the exam.

What two main problems do you see?

> "The process of elimination, combined with a modicum of common sense, will always assist us to arrive at the correct conclusion with the maximum of possible accuracy and the minimum of hard labor. Which being translated means: I guessed it."
>
> —Margery Allingham's detective, Albert Campion, in *Look to the Lady*[xxix]

Ocean has a distended abdomen with nausea AND skin issues.

These are probably two completely separate problems.

Maybe you should make two differential lists…

Differential Diagnoses (distended abdomen and nausea):

- Congenital disease — An abnormality a patient is born with.
- Gastric overdistension — A stomach so swollen it hurts and doesn't function correctly.
- GI foreign body
- GI parasitism
- Infection
- Toxin
- Trauma — Blunt trauma to the abdomen can cause things like a ruptured bladder or internal bleeding that would distend the belly.

Differential Diagnoses (hair thinning, itching, and dandruff):

- External parasites (like mites, lice, or fleas)
- Poor diet
- Ringworm — Not a worm, but a fungal infection of the skin.
- Seborrhea — A disorder of skin keratinization (the normal process in which skin cells get horny from stuff called keratin and are shed away).
- Skin infection

Decision time!

What tests do you want to run first?

Here's what you can do in your clinic:

- ✓ Basic bloodwork (a chemistry panel and CBC)
- ✓ Blood gas analysis
- ✓ Clotting times
- ✓ Cytology
- ✓ Fecal floatation and direct saline exam
- ✓ FeLV/FIV test
- ✓ Giardia test
- ✓ Heartworm test (with rickettsial diseases)
- ✓ Leptospirosis test
- ✓ Pancreatic lipase (fPL or cPL)
- ✓ Parvo test
- ✓ Pregnancy test
- ✓ Radiographs (X-rays)
- ✓ Skin scraping
- ✓ Thyroid hormone level (T4)
- ✓ Ultrasound
- ✓ Urinalysis
- ✓ Wood's test or DMT plate

What tests do you want to run?

You recommend a **parvo test first, then a radiograph and fecal** (and maybe a brief **ultrasound** to look for free fluid in the abdomen). You also recommend **skin scraping** for the skin issues you noticed.

"Well…" she scratches her leg and looks at the ceiling for several minutes until you think she might have dozed off. But whatever invisible apparition she was interested in must have faded, and she finishes her sentence. "My boyfriend just lost his job, and we can't afford much. He's a chef, so at least he's home cooking for me now." She giggles, and you smile a bit. But you still need an answer. In the end, she chooses from an estimate, declining a fecal test and ultrasound, but approving the other diagnostics.

You run a parvo test and it is *negative*. Woote!

You take a radiograph and see Something that could explain poor Ocean's GI signs right away. And since Ember's asked you to keep the cost down you tell your techs to stop with the one view.

Here it is (with a couple of markers to help you get your bearings…)

Try to answer this before looking at the next page.

What's going on with Ocean's abdomen?

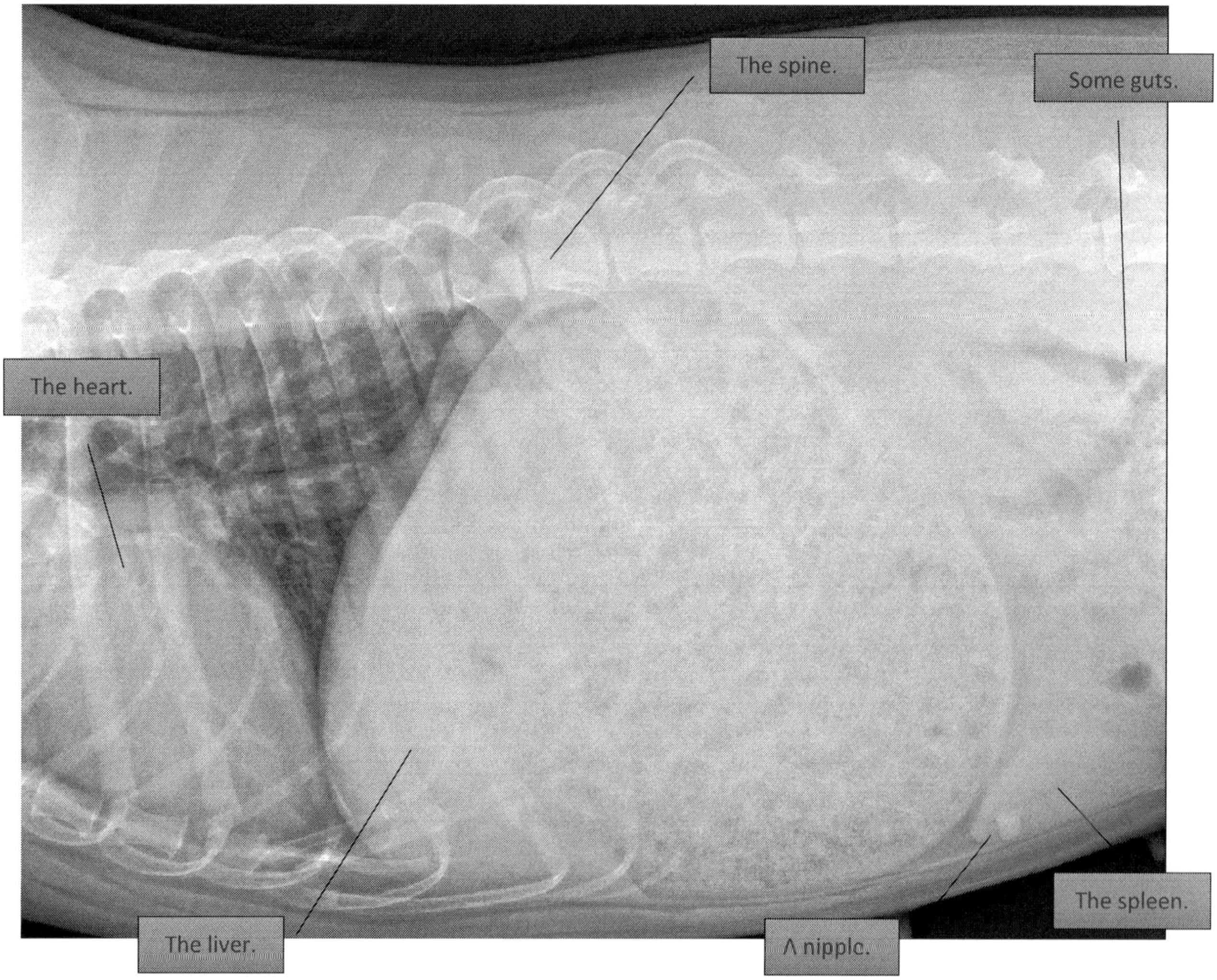

That big, big round thing without a label (sitting in between the liver and the spleen) is Ocean's stomach. It's very distended and filled with "gastric material" that looks very much like an excessive amount of food. Technically, you can't be sure it's food. Cloth, foreign material, or some kind of weird mass could look like that too. But given this patient's history and signalment, it's probably food…

But what about his skin scraping?

You have Bojka wear some gloves and hold the squiggly puppy to scrape his skin with a dull blade, applying your sample obtained this way to a drop of mineral oil on a slide. You also employ the "acetate tape technique" to get a sample of the strange moving dandruff. This means you take a little piece of scotch tape, press it onto an affected area of fur a couple of times, then put it over a drop of mineral oil on a slide.

You take a look in the microscope. Out of the two samples, you see much more interesting stuff with your tape technique… And you immediately start feeling itchy.

Here's a glimpse in the microscope:

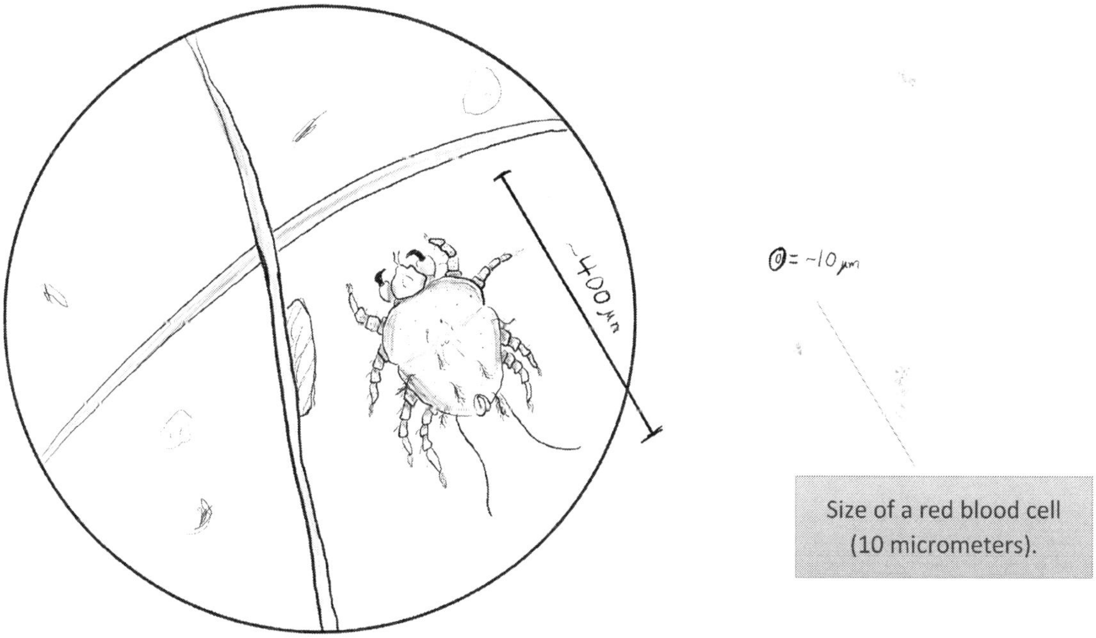

You bravely resist the urge to leap away from the microscope and scream, "Gaaaaaah! It's horrible! It's horrible!" Instead, you calmly estimate this alien creature crawling across the field of view to be about 400 micrometers long. For comparison, a red blood cell is about 10 micrometers. What's going on here?

And your diagnosis for both conditions is?

STOP!!!!!!

OK… It's hard to stop scratching your arms and legs and face and head now. Or think at all about that radiograph. But let's go over it first, since poor Ocean's belly seems to bother him the most.

You diagnose **gastric overdistension**, also commonly called "food bloat." This is not the same condition as GDV (gastric dilatation and volvulus) which is also called "bloat." In GDV the stomach fills with gas and flips over. In "food bloat" the stomach gets overfilled with food to a point where it's hard to even empty like it normally does. It's not rotated ("torsed"), just super-super full.

This is often something you see in young, greedy puppies who've gotten into their pal's food or snuck into a bag of dog food. They'll have signs of abdominal distension and discomfort, as well as nausea. Usually, they can't actually vomit, because their stomach is so very distended it can't contract like it normally does. Gas and diarrhea will develop as the food sloooooowly moves through their intestines. Slowly is the operative word. A thing called ileus becomes a problem in puppies with gastric overdistension. Their intestinal tract moves stuff through slower than normal, basically because it's overwhelmed.

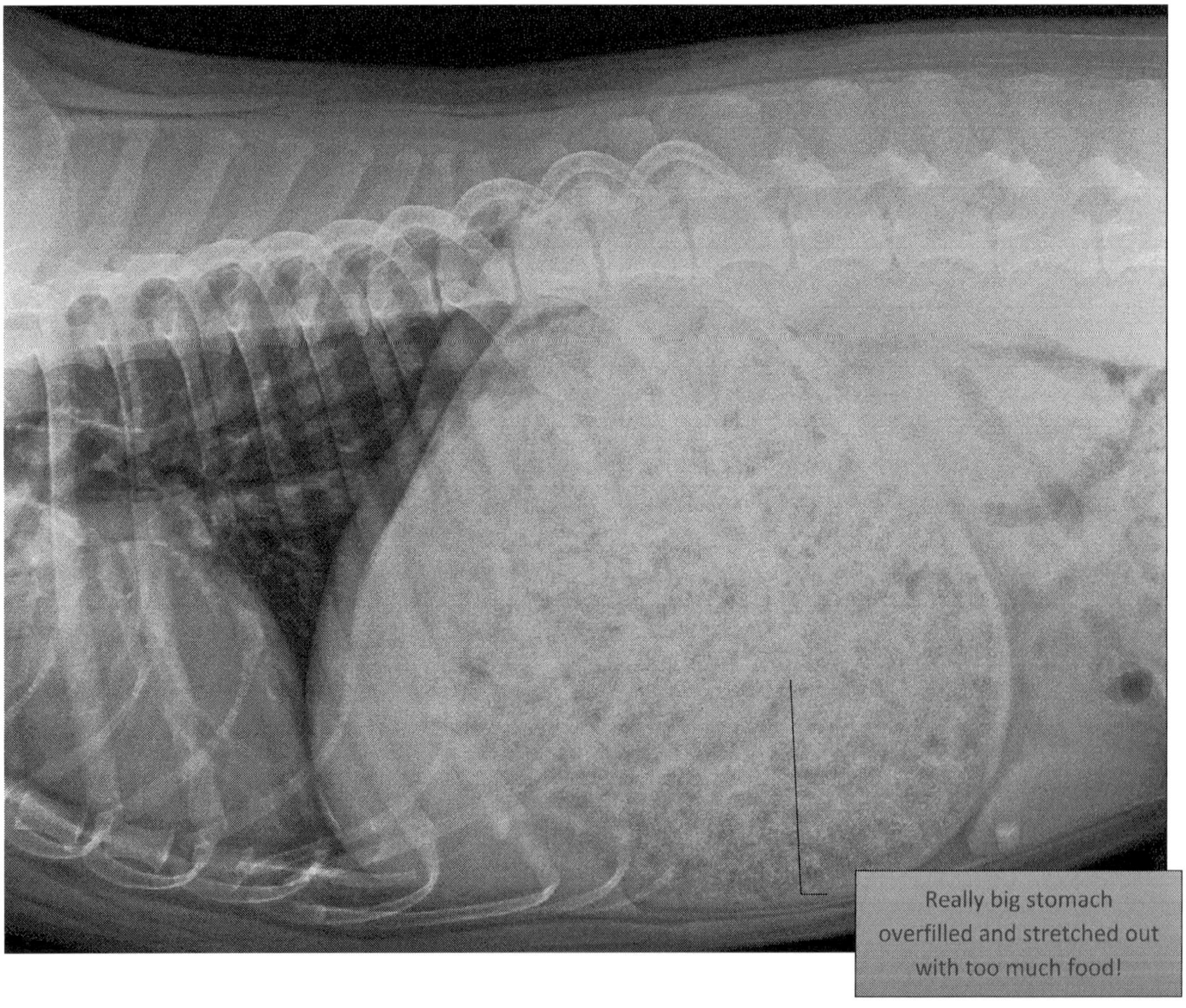

On to the skin scraping! You recognize that horrid little microscopic monster. It's a *Cheyletiella* mite, just like you feared. These teeny bugs will infest a dog's skin and tunnel under the surface, making little flakes that look like dandruff and that sometimes can be seen moving. Usually, you see signs of itching, flaking, and "walking dandruff" along an infested dog's back with this particular mite.

Unlike other types of mange (the word for mite infestation in pets), *Cheyletiella* species are zoonotic. This means they can affect people too. They chew up human skin and make red, itchy rashes. You strongly suspect that's why Miss Daydream is scratching herself all over your exam room. No fun!

There are several species of *Cheyletiella* mites, but the most common type living on dogs is *C. yasguri.* It lays eggs on its host and in the environment and has an average life cycle of 21 days. There's an egg visible on one of the hairs you saw in the microscope...

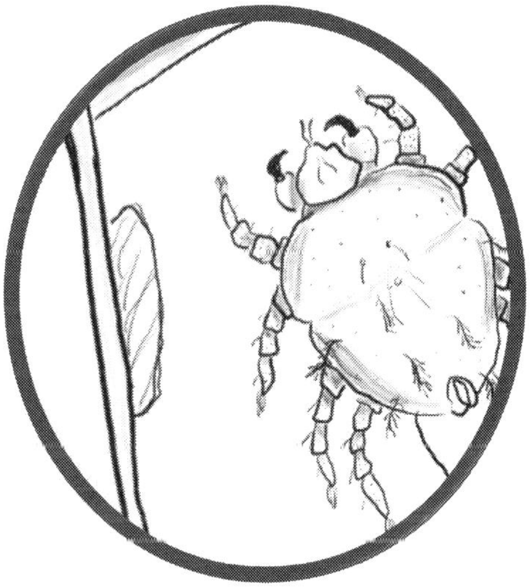

Oh boy... Somebody should really do something about all this!

Oh, right. That's you. You're the doctor.

You should do something about it.

What's your treatment plan?

Mange: mite infestation in pets.

Zoonotic: diseases that can be transmitted from animals to humans.

Pick some treatments

a) Abdominal surgery

b) Antibiotics

c) Deworming medication

d) Flea and tick medication

e) Fluids (IV or SQ?)

f) Happy vibes

g) A laxative

h) Lime sulfur dip

i) Nausea drugs

j) Prescription food

k) Probiotics

l) Prokinetics

m) Steroids

> Medication that makes a patient poop.

> A sulfurated lime (which smells like a convention of skunks) is often used for fungal and parasitic skin conditions.

> Medication that encourages movement of the intestines.

You recommend **d, e (sq), i, and l.**

Subcutaneous fluids. Fluids will help with Ocean's hydration status and promote digestion of all that food in the stomach.

Nausea drugs and prokinetics. You'll also give a nausea injection and prescribe a nausea/prokinetic medication called metoclopramide. Some cases of gastric overdistension benefit from hospitalization and IV fluids, but this is a case where cost is of concern, and Ember appreciates the less expensive option.

"And how about those…" she stops and covers Ocean's ears. "…those bugs you say are living on Ocean?"

"Fortunately, they're susceptible to several different topical **flea medications** (d). Let's use one of those!"

"Why'd you say (d)?"

"Oh, sorry. Just ignore that."

You decide against lime sulfur dip even though that would also be an acceptable treatment for *Cheyletiellosis.* It is super stinky (as in, your clinic would smell like an army of skunks had mobbed it for about a week kind of stinky). It's also something she'd need to come back for multiple times. There are severe cases where oral medications are given for these mites. Those can have some serious side effects, and you wisely advise the simple option first.

"These critters lay microscopic eggs in your house and yard too, so you should talk to an exterminator. There are some sprays you can buy at stores to kill this mite. Do you have any other pets or people in your home that could be infested?"

Miss Daydream's aura sags. "I have two other dogs, three cats, a racoon, a rabbit, and a monkey… Oh, and three children too. They *have* been acting awfully itchy lately."

"Er… which ones?"

"All of them, really."

"Ah. Well, I'd recommend you consult your pediatrician and physician." You almost add "and maybe the army and National Guard," but manage to stop yourself. "And your other dogs and cats and rabbit are likely to be infested. Why don't you bring them in to get weights and we'll treat them all? The raccoon and monkey aren't precisely legal to own here in New Mexico. Maybe you have some kind of special permit for them? Where'd you get 'em, can I ask?"

She gives you a sheepish smile. "Oh…um… Coony came from the forest as a baby. Monkey Man came from…a guy." Then she starts scratching her arms and neck and nose.

"A guy?"

"Mmm hmmm. He's, you know, local." She is obviously not interested in saying more, and as she takes off her shoes to scratch her feet, you decide to get her on her way. Afterward, you're very tempted to give yourself and the entire clinic one of those lime sulfur dips. And maybe then torch the place and start all over from scratch.

Instead, you try to think about something else. And that's easier to do, now that some guy is yelling in the lobby…

CASE #10

"Mr. Hatahle! Please calm down…" You hear Amy Wallender pleading as you hurry toward the lobby.

"You did something to Lobo! He's all swollen! I need to see a doctor now!"

You find James Hatahle, a regular client, looking furious in the lobby. He's holding a familiar Labrador puppy, with an obviously swollen face.

"Hello, Mr. Hatahle. What's going on?"

He turns to fume at you now, but you've gallantly rescued Ms. Wallender. "I took Lobo home and went to eat at La Chuleta, home of Santa Fe's most delicious chile rellenos (though sadly home to very few customers). When I came back, I found my Lobo's face all swollen like a desert toad! And he vomited on the way here too. What was in that shot you gave him?"

You take a quick look at Lobo and assure his owner he looks stable. Then you convince him to follow you and Fred into an exam room where you can do a more thorough exam.

Step 1 – Organize the History

Signalment:

4 month old

Labrador

Male intact

Presenting Complaints:

Facial swelling

Vomiting

Step 2 – Gather Clues

Physical Exam:

Abdominal palpation: WNL, no enlarged organs or masses noted.

Attitude: BAR.

Cardiac: No murmur or arrhythmia noted, femoral pulses are strong.

Ears: Bright pink color to skin inside pinnae (the official name for the pointy, floppy part of an ear).

Eyes: A little bloodshot bilaterally.

Integument: Swelling noted of upper lips and around both eyes. Multiple raised, rounded, firm lesions noted over the patient's head, neck, and trunk.

Lymph Nodes: Normal sizes.

Musculoskeletal: WNL.

Neurologic: WNL.

Oronasal: Teeth are clean. No nasal discharge noted.

Pulmonary: Lungs sound clear, no increase in respiratory effort noted

Urogenital: WNL.

Vitals: MM pink and moist, CRT <2 seconds.

Differential Diagnoses:

Hypersensitivity (allergic) reaction

Insect stings

Skin infection

Trauma

Snakebite

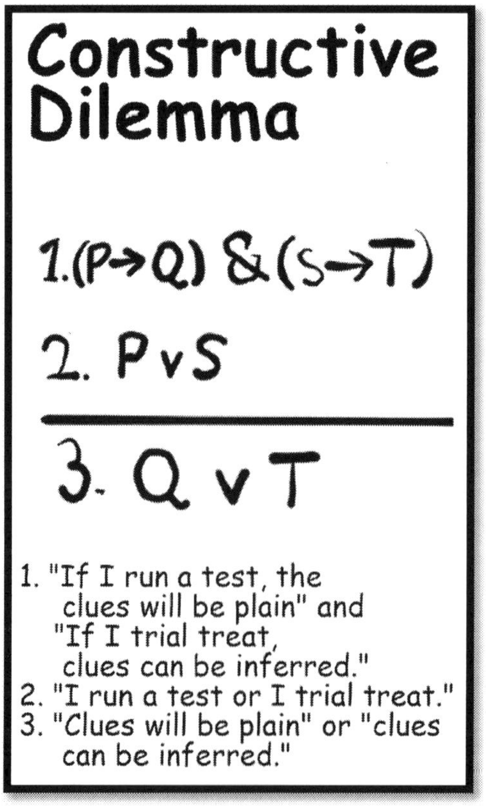

Decision time!

What tests do you want to run first?

Here's what you can do in your clinic:

- ✓ Basic bloodwork (a chemistry panel and CBC)
- ✓ Blood gas analysis
- ✓ Clotting times
- ✓ Cytology
- ✓ Fecal floatation and direct saline exam
- ✓ FeLV/FIV test
- ✓ Giardia test
- ✓ Heartworm test (with rickettsial diseases)
- ✓ Leptospirosis test
- ✓ Pancreatic lipase (fPL or cPL)
- ✓ Parvo test
- ✓ Pregnancy test
- ✓ Radiographs (X-rays)
- ✓ Skin scraping
- ✓ Thyroid hormone level (T4)
- ✓ Ultrasound
- ✓ Urinalysis
- ✓ Wood's test or DMT plate

What tests do you want to run?

This is a bit of a "trick question." While it wouldn't be wrong to run bloodwork, it's unlikely to show anything useful in this case. One test that might help would be a quick ultrasound of the abdomen, especially to look at the gallbladder. We'll talk about why in a minute… This time you decide to do no tests.

And your diagnosis is?

STOP!!!!!!

That's right! This is a **hypersensitivity reaction**, also called an allergic reaction ("allergic" is from a Greek word indicating "altered reaction").

This happens when the body's immune system overreacts to a molecule that is not actually harmful. We call the molecule that triggers a hypersensitivity reaction an "allergen." Allergic reactions happen because the body has seen the allergen sometime in the past and mounted an immune response that created a memory of that antigen as a nasty, foreign thing.

Many things can trigger allergic reactions, including insect stings, environmental allergens (like pollen or dust mites), drugs, or vaccines. Vaccines are a fairly common reason for allergic reactions in dogs. It's usually the adjuvant (chemical preservatives) in the vaccine that'll cause a reaction.

There are four basic types of hypersensitivity reactions. Most of them happen through the action of specific types of antibodies (immunoglobulins), which are molecules created to fight a specific antigen. Seasonal allergies (that runny nose and itchiness you may be cursed with every year) are a type I hypersensitivity. So are most reactions to vaccines.

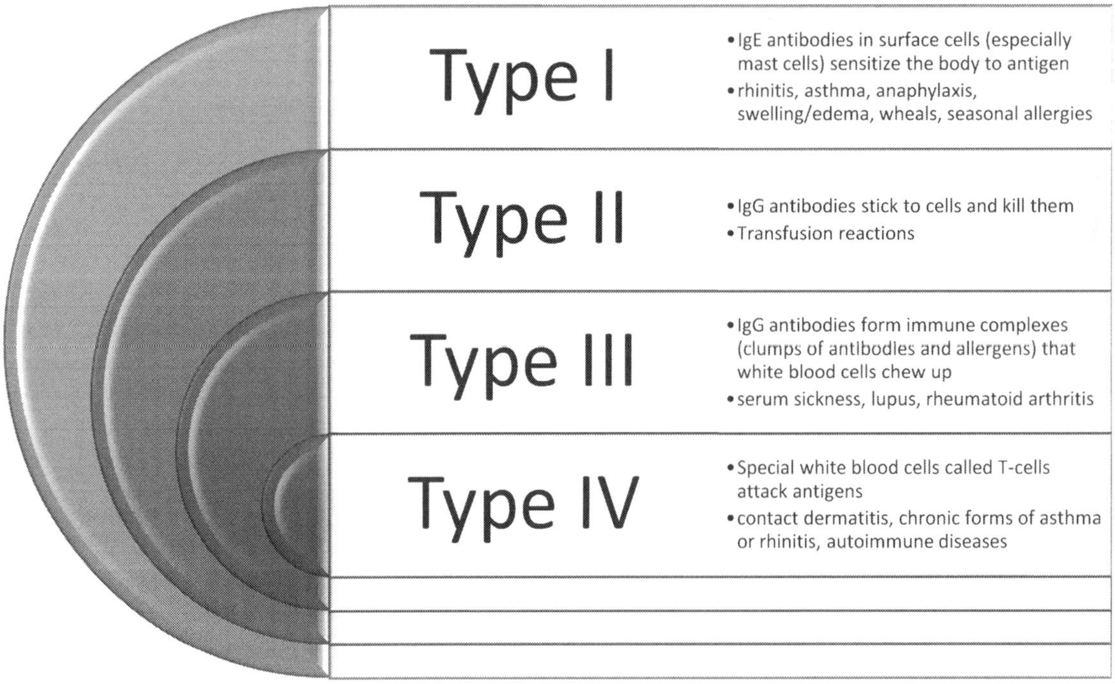

Is a type I hypersensitivity reaction to a vaccine deadly? It depends. Usually not. There are rare cases of anaphylactic reactions (systemic, or whole-body reactions) that can be severe enough to be fatal.

Allergen: a harmless molecule that triggers the immune system into attacking it.

Anaphylaxis: a serious systemic or whole body reaction to an allergen.

Antibody: molecules the immune system forms to fight allergens (and real infections).

Antigen: a "foreign" molecule the immune system makes an antibody to.

Hypersensitivity reaction: an unnecessary or overreactive immune system response to an allergen.

Often, the only way to know where a patient is in the spectrum of mild hypersensitivity to severe anaphylactic reactions is by guesswork based on their clinical signs and vitals. One useful trick in dogs, in particular, is to look for a "halo" sign on their gallbladder with your ultrasound machine. This looks like a bright ring around the gallbladder and is a sign of inflammation that often occurs with canine anaphylaxis (as opposed to more contained hypersensitivity reactions). This sign might occur with heart problems, too.
"So you're telling me the vaccine that *you* gave has made Lobo all bumpy and swell?"

You wisely repress the urge to giggle at the words "bumpy and swell." You explain to him that, yes, the vaccine probably caused this reaction. But it's not your fault and has nothing to do with the way you gave or handled the vaccine. It's probably not even the vaccine manufacturer's fault. They make all the vaccines the same and need to use things like adjuvant in every vaccine or they wouldn't be safe to store. It's something that could happen to any dog, and just can't be predicted.

"Hmm…" James Hatahle does not look convinced. "The vaccine was given here. Lobo had a reaction to that vaccine. So you should make it right, whatever it takes."

You calm James down by assuring him that Lobo is quite stable and there's every reason to think that a few simple treatments will fix him.

Naturally, he asks what the treatment is. And naturally, you tell him.

Choose from this list of treatment options.

a) Antibiotics
b) Antibiotic oral medications
c) Antihistamine injection — A drug that blocks the release of histamine (a chemical causing inflammation in allergies).
d) Antihistamine oral medication
e) Deworming medication
f) Epinephrine shot — An important hormone splurted from the adrenal gland in response to stress.
g) IV fluids
h) Jump around in circles and scream like a baby yak — How does a baby yak scream?
i) Nausea drugs
j) Monkey spittle
k) Prescription food
l) Steroid injection
m) Steroid oral medication
n) Surgical removal of hives
o) SQ fluids

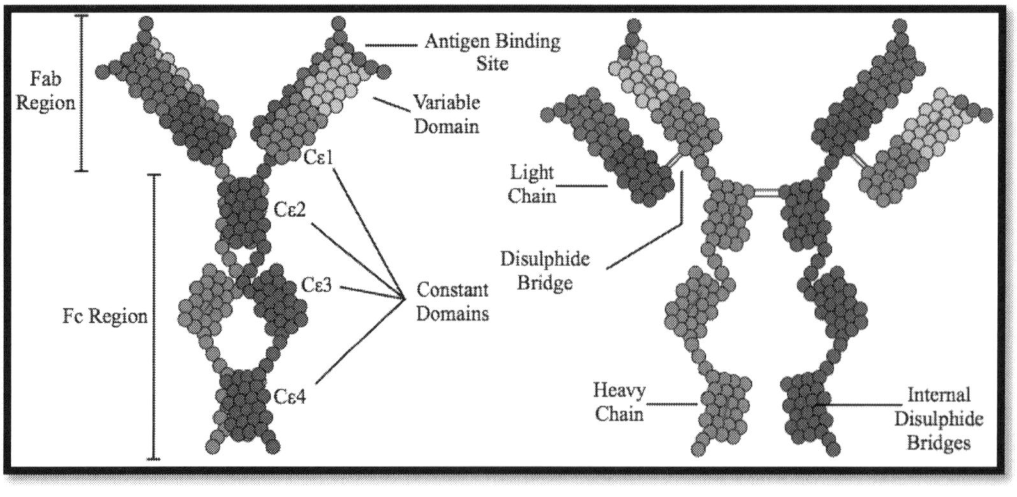

The structure of an Immunoglobulin E (IgE) antibody molecule

Yes! Good choice picking **c, d, i, and l.**

Steroid injection. You tell Mr. Hatahle that you need to give a steroid injection to fight the immune system's overreaction and reduce swelling.

Antihistamine injection and medication. You'll also give a diphenhydramine injection, which is a type of antihistamine for dampening the response of mast cells in allergies (mast cells are big cells that live in the skin and dump inflammatory chemicals out in response to allergens). You'll also calculate a dose of oral diphenhydramine for him at home until the swelling goes down completely.

Nausea. Since he vomited, you'll also give an injection for nausea.

That's all there is to it! Bing, bang, boom!

For more severe reactions, you sometimes give epinephrine (especially in serious anaphylactic reactions). Anaphylaxis cases often are in shock and need IV fluids and intensive monitoring and treatments (for things like low blood pressure). Sometimes you send home oral steroids for more chronic allergy issues.

You write some doses and Fred and Bojka pop some injections into little Lobo (lest the great and powerful doctor soil his lily hands). A few minutes later, Lobo's looking a bit better. He's still very stable with normal vital signs, and you send him home.

James Hatahle shakes your hand and stares soberly into your eyes. Your white man imagination hears an eagle screech and a mournful flute echo somewhere. "Thank you. You have done the right thing and are appreciated." He apologizes to Ms. Wallender for his earlier panic and you all start feeling better.

Until two old people stumble in carrying a seriously weak shepherd cross who can't walk. She's crying and he's panicked. You recognize one of your oldest and best clients: Hector and Susan Melendez…

Case #11

"*¡Ayudame!*"[10] Mr. Melendez calls out as you and Fred hurry over to take the dog.

"Raaaaar! Raaaargh! Arf! Snarl!" the dog calls out, as she tries to rip off a hand from each of you.

"Muzzle!" you call out, and Bojka rushes in with a handy nose-sock to help keep you all safe. Fortunately, the patient is weak enough that she doesn't put up much of a fight after her initial barrage of teeth. As you put on the muzzle, you notice her gums are very pale. You carry the shepherd to the treatment area in the back and motion for Mr. and Mrs. Melendez to follow. You ask questions as you do a cursory examination.

"Tell me what's happened."

Mrs. Melendez speaks up. She's in a trim business suit, and you recall she works as a realtor. "I came by the house this morning to pick up some paperwork for the business and found Flower laying on her side. She seemed very weak and was breathing heavily."

You agree with that assessment. The patient seems unable to stand on her own, though she can sit sternal. She is breathing very fast, and it seems like she has increased effort on expiration. Her lungs sound harsh everywhere, but you can't really hear any specific crackles or wheezes.

"I called Hector at the restaurant, and he rushed home to help me bring her to you."

"*Afortunamente, estamos non muy ocupado allí*,"[11] Mr. Melendez adds. You recall from past visits that he understands English very well, but is still learning to speak it… Much like you can understand a lot of Spanish but struggle to speak it.[12]

You notice Flower's femoral pulses are very weak and tell Fred and Bojka to set up an IV catheter.

"Have you noticed anything unusual before this? Any coughing, sneezing, vomiting, or diarrhea?" You remember the very literal Bernard from yesterday and steel yourself to hear more information than you want about your clients. But fortunately, they understand you're talking about Flower.

Expiration: one of the two phases of breathing. You inspire when you breathe in, you expire when you breathe out. You also expire when you die, but that's a whole different synonym, baby!
Sternal: having to do with the side of the body near the sternum (the "breastbone"). Often we use this word to mean a dog can sit up enough to lie on her chest.

---

[10] "Help me!"
[11] "Fortunately, we are not very busy there."
[12] The last time, in fact, that you tried to have a conversation in Spanish with a client she ended up asking you how much a seizure costs and whether it comes in pink. You'd been proud of how you were doing up to that point. After that, you decided to start online Spanish courses!

"No. She's been acting fine. Nothing like that."

"A little coughing," Hector adds.

"True. She's coughed on and off infrequently for the past month or so. We didn't think much of it."

You notice pale scars on the patient's neck and flanks. They create patchy areas where the hair isn't grown well. You ask them about this.

"Flower came from our son about two years ago," Susan explains. "She had those scars from getting into fights with his dog. That's why Arturo sent her to live with us." Mr. Melendez adds something in Spanish, and she translates. "Yes… He told us a coyote or wolf attacked her more than once."

There are an awful lot of scars, but they're not what's killing her now, so you nod and move on to a finding of more immediate concern. It looks to you like Flower's abdomen is distended. You palpate it, and it feels like there's fluid inside…and maybe a firm something in the cranial abdomen. You're getting an idea of what you're dealing with here.

This patient needs help. You can find out more history details later. It's time to organize your history and PE and get started!

Step 1 – Organize the History

Signalment:

11 year old

Shepherd mix

Female, spayed

"Facts don't care about your feelings."
—Ben Shapiro[xxx]

Presenting Complaints:

Acute collapse

Dyspnea

Weakness

Step 2 – Gather Clues

Physical Exam:

Abdominal palpation: Distension noted, with a fluid wave suspected. Suspect possible cranial abdominal mass (or enlarged spleen or liver) on palpation as well.

Attitude: Quiet, weak, but responsive.

Ears: Appear clean bilaterally.

Eyes: No significant findings. Appear clear.

Cardiac: Possible grade 2/6 murmur noted on the left (it's hard to be sure with the increased lung sounds drowning out heart auscultation). Weak femoral pulses bilaterally.

Integument: Patchy alopecia with multiple scars over the neck, shoulders, and flanks.

Lymph Nodes: WNL.

Musculoskeletal: No significant findings. You don't notice any palpable fractures or luxations, and she doesn't act painful when you flex and extend any of her joints. She seems comfortable on spinal palpation.

Neurologic: Weak and somewhat listless, but responsive. Normal menace reflex and palpebral reflex bilaterally. Normal PLR bilaterally.

Oronasal: No significant findings.

Pulmonary: Moderate dyspnea (increased expiratory effort and tachypnea, with no cyanosis noted). Harsh lung sounds in general, with no specific crackles or wheezes.

Urogenital: No significant findings.

Vitals: Pale MM, cannot assess CRT (because P wants to eat you).

Alrighty. At this point in the case, you probably have a reasonable differential list forming in your head, which will help you recommend some tests to run. And by this time in this book, you should be able to form some differentials on your own.

Write all the possibilities you can think of that would cause Flower's signs.

Make your own differential list, then turn the page to compare it.

Alopecia: hair loss.
Auscultation: listening with a stethoscope.

Differential Diagnoses:

- Cardiogenic shock
  - Low blood pressure from heart failing.
- Congestive heart failure
- GD
  - GD=Gastric dilatation: a gas-distended stomach.
- GDV
  - GDV=Gastric dilatation and volvulus: a gas-distended stomach that's flipped and twisted.
- Hepatic failure
- Hypovolemic shock
  - Whole-body drop-in blood pressure in response to a disease, starving the tissues of oxygen.
- Immunologic disease (like IMHA)
  - Immune mediated hemolytic anemia: the body attacking its own blood.
- Intervertebral disc disease (IVDD)
  - A "slipped disc" in the back, putting pressure on the spinal cord that can paralyze a patient.
- Neck injury
- Neoplasia (cancer)
- Stroke
- Trauma

"Seeing may be believing, but, in order to understand what you are seeing, logic is necessary."

—George Hammond[xxxi]

Decision time!

What tests do you want to run first?

Here's what you can do in your clinic:

- ✓ Basic bloodwork (a chemistry panel and CBC)
- ✓ Blood gas analysis
- ✓ Clotting times
- ✓ Cytology
- ✓ Fecal floatation and direct saline exam
- ✓ FeLV/FIV test
- ✓ Giardia test
- ✓ Heartworm test (with rickettsial diseases)
- ✓ Leptospirosis test
- ✓ Pancreatic lipase (fPL or cPL)
- ✓ Parvo test
- ✓ Pregnancy test
- ✓ Radiographs (X-rays)
- ✓ Skin scraping
- ✓ Thyroid hormone level (T4)
- ✓ Ultrasound
- ✓ Urinalysis
- ✓ Wood's test or DMT plate

What tests do you want to run?

You recommend general **bloodwork**, as well as doing a brief **ultrasound** of your patient and probably **radiographs** (of the chest and abdomen). You explain you suspect fluid in the abdomen on physical examination and will look for that in particular on ultrasound of the abdomen. If there's enough fluid, you'd like permission to schluck some out with a needle (a procedure called abdominocentesis… Because everything in medicine needs a big fancy name). The Melendez family approves.

You and Fred start diagnostics. You draw some blood and get it running. Then you do an ultrasound of the abdomen. You look at a specific set of views you've trained to focus on in emergencies, and you find free fluid in three of them.

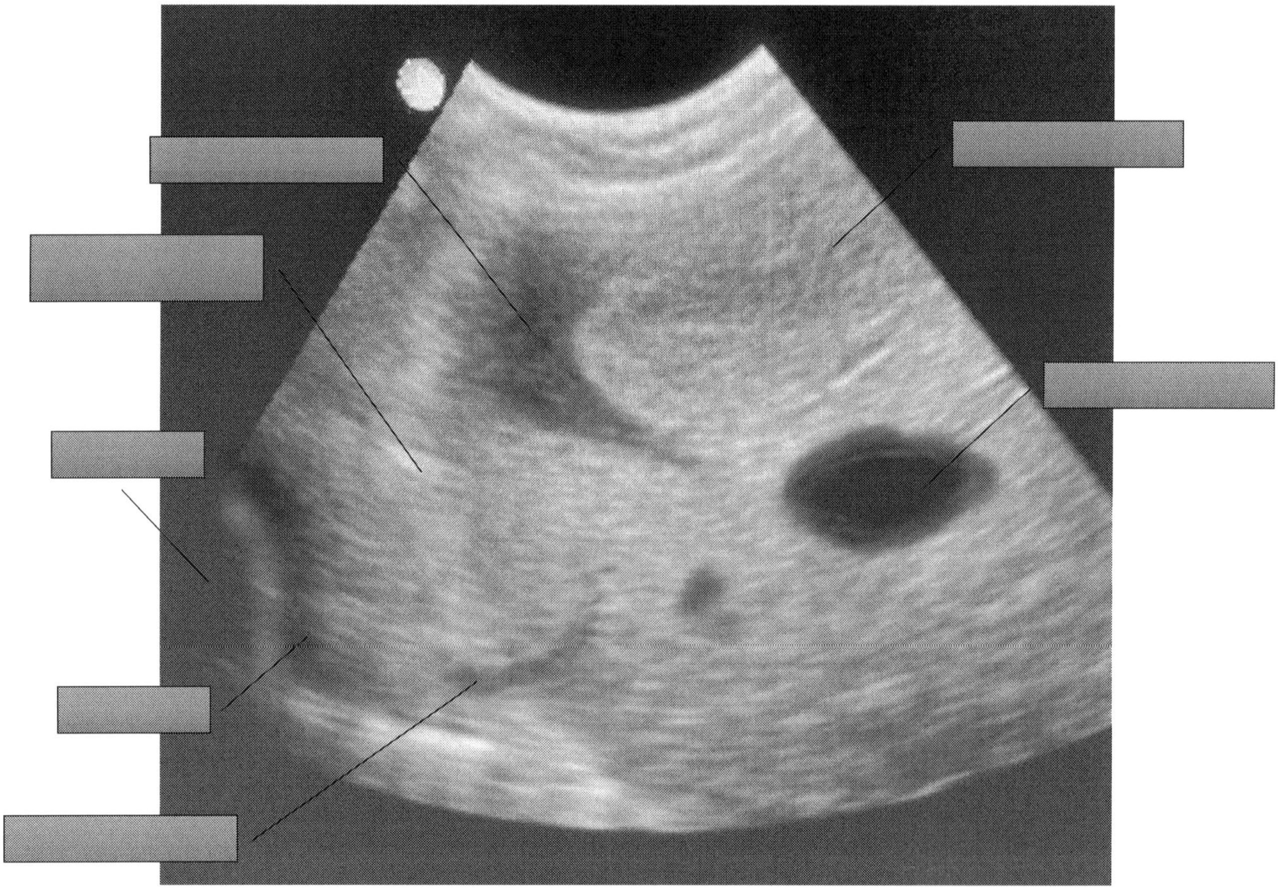

Here is one of the more important views, usually called the diaphragmatic-hepatic view. It's seen by setting the ultrasound probe against the very front of the abdomen up by the xyphoid (the tip of the sternum) and directing it a bit toward the diaphragm and chest cavity (which lays in front of the diaphragm).

Any idea what the structures you're looking at are?

Abdominal free fluid: abnormal fluid, floating around inside the abdominal cavity. Also called ascites.
Xyphoid: the tip of the sternum.

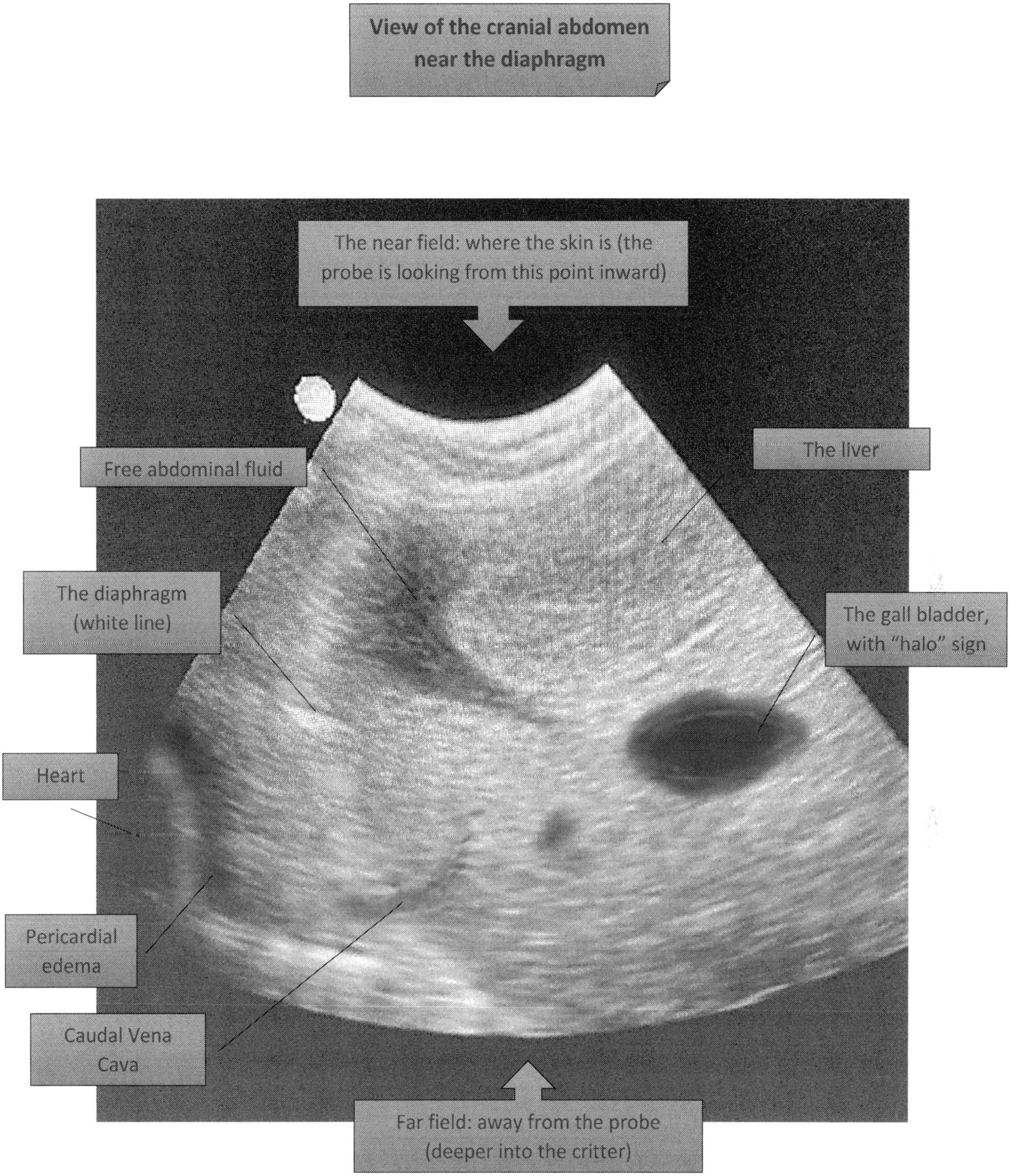

Caudal vena cava: a big blood vessel that brings blood from the body back into the right side of the heart. It goes through the diaphragm and can be seen on ultrasound.
Pericardial edema: fluid building up in the sack around the heart (the pericardium).

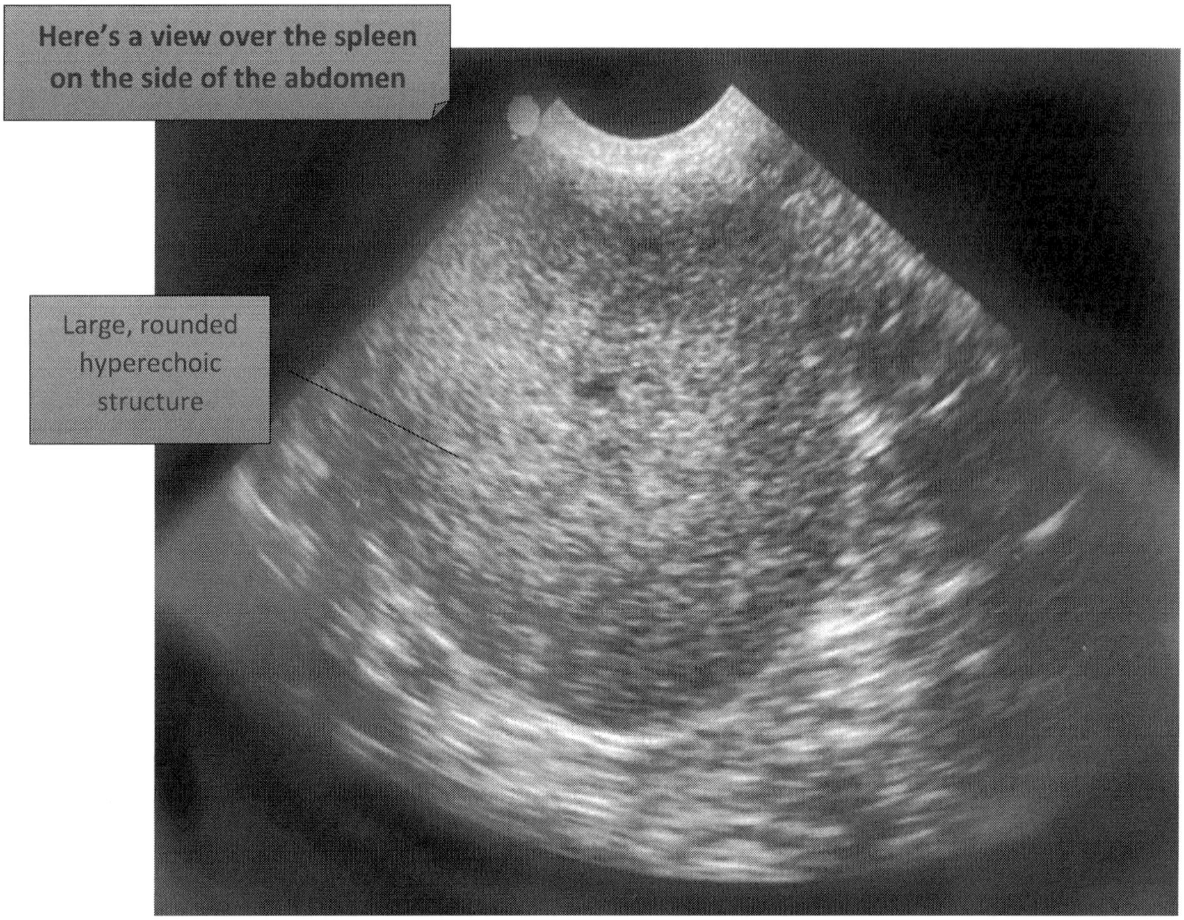
Here's a view over the spleen on the side of the abdomen
Large, rounded hyperechoic structure

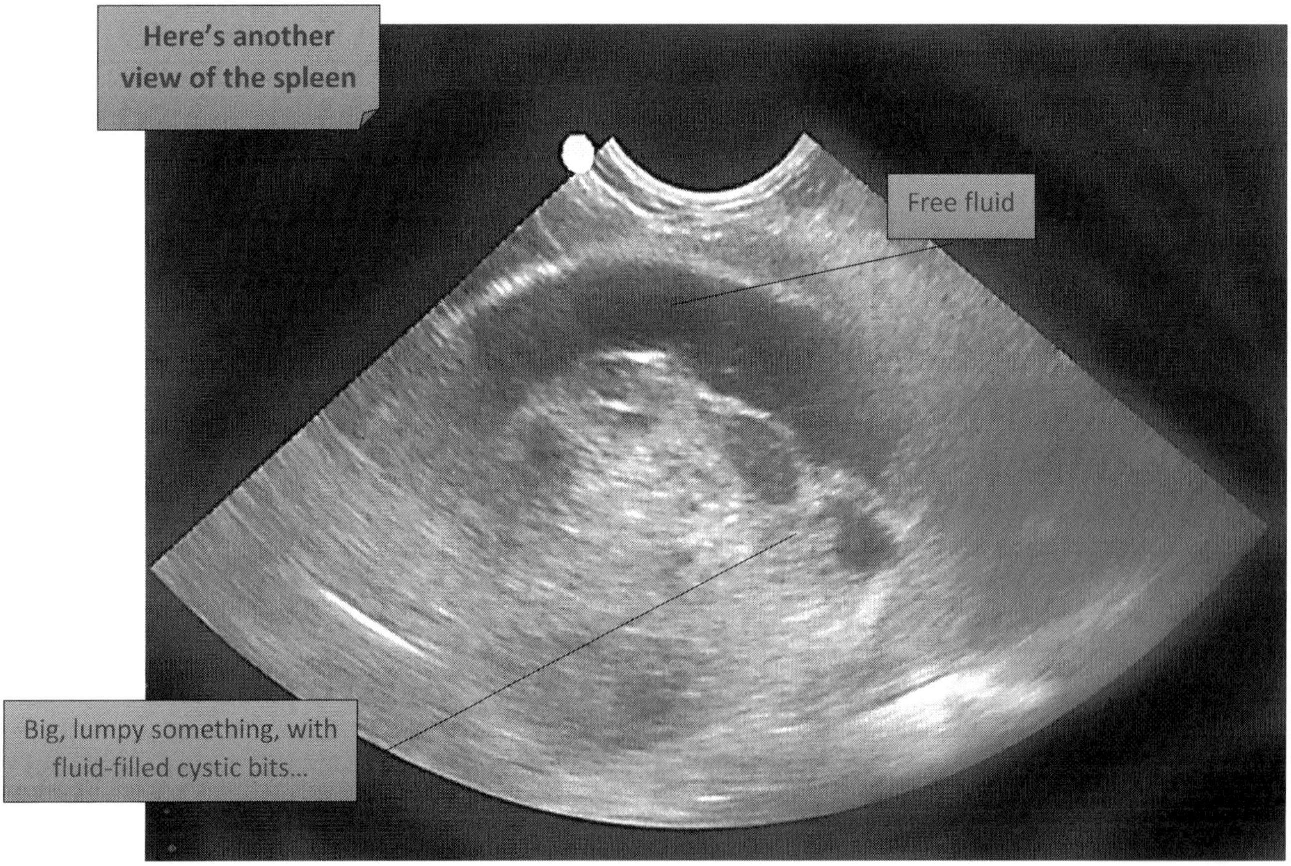
Here's another view of the spleen
Free fluid
Big, lumpy something, with fluid-filled cystic bits...

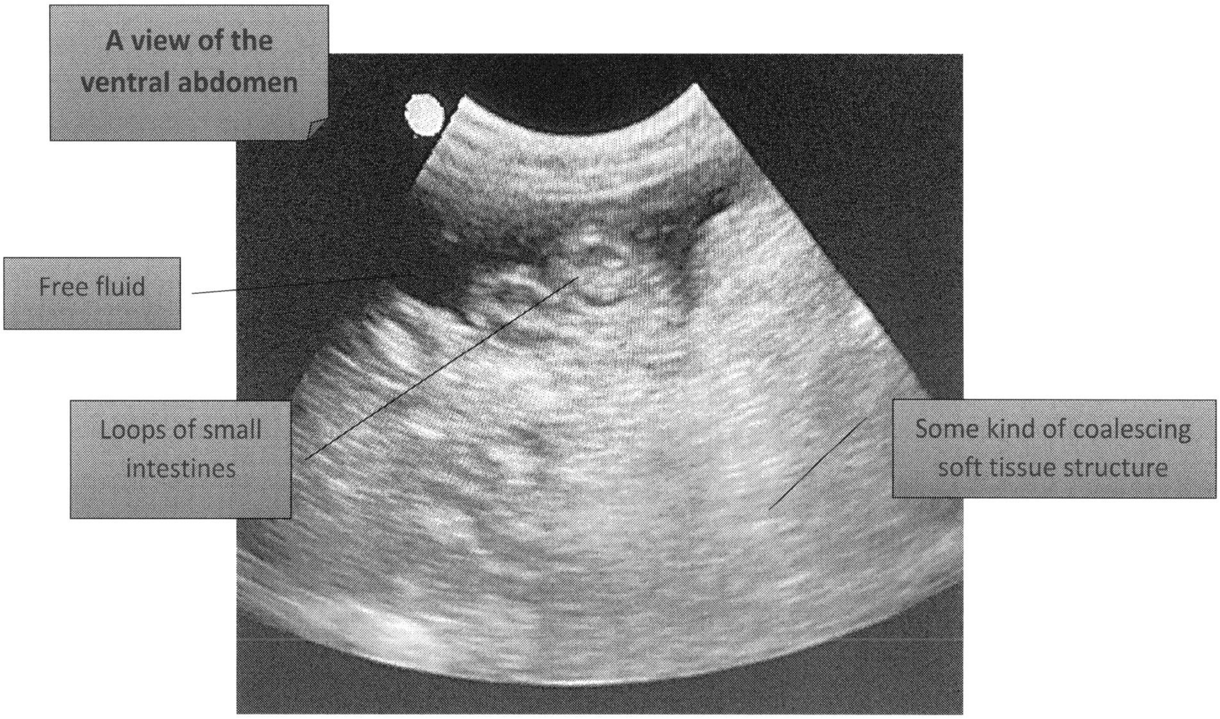

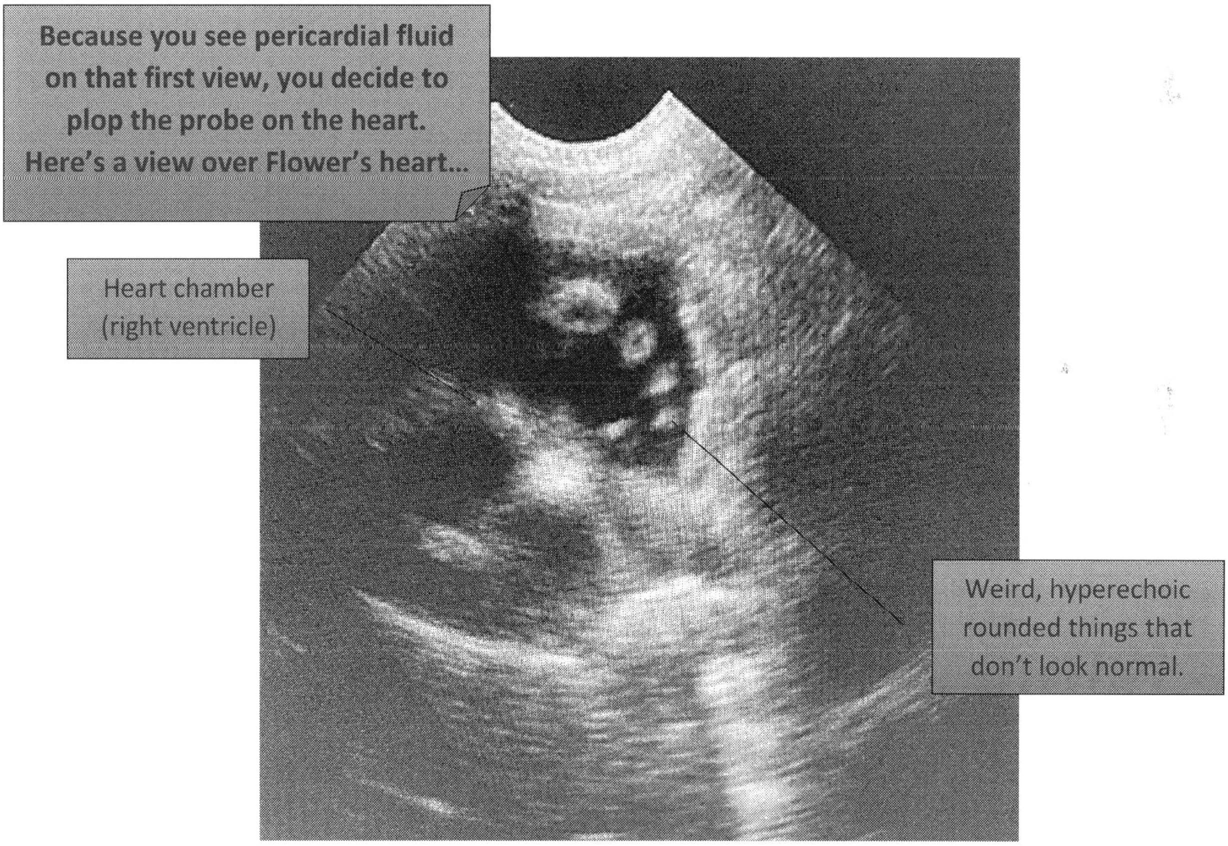

You sample some of that free fluid in the abdomen with a needle. It looks like a bunch of blood. Hmm… Things don't look good so far.

At this point, you may have enough information for a diagnosis, but you go ahead and do your radiographs and let the bloodwork finish, just to confirm your dire suspicions.

Client: Melendez, Hector	Species: Canine	Sex: FS
Patient: Flower	Breed: German Shepherd cross	
Patient ID: 026601		

CBC

TEST	RESULT	FLAG	REFERENCE RANGE[xxxii]	
WBC	10		6.0-17.0	10x3/µL
Neutrophils	5		3-11.4	10x3/µL
Monocytes	1.1		0.15-1.35	10x3/µL
Lymphocytes	3.1		1-4.8	10x3/µL
Eosinophils	0.4		0.1-0.75	10x3/µL
Hematocrit	19%	L	37-55	%
MCV	67		60-77	fL
Platelet Count	205		200-900	10x3/µL

Client: Melendez, Hector	Species: Canine	Sex: FS
Patient: Flower	Breed: German Shepherd cross	
Patient ID: 026601		

Chemistry Panel

TEST	RESULT	FLAG	REFERENCE RANGE[xxxiii]
ALT	48		8.2-57 IU/L
ALP	98		10.6-101 IU/L
GGT	6		1.0-9.7 IU/L
tBili	0.4		0.1-0.6 mg/dL
Glucose	100		62-108 mg/dL
Cholesterol	221		116-254 mg/dL
Total Protein	3.5	L	5.5-7.5 g/dL
Albumin	2.8		2.6-4.0 g/dL
Globulin	3.1		2.1-3.7 g/dL
Calcium	11		8.7-11.8 mg/dL
BUN	48	H	8.8-26 mg/dL
Creatinine	8.7	H	0.5-1.6 mg/dL
Phosphorus	12	H	2.9-6.2 mg/dL
Sodium	145		140-154 mEq/L
Potassium	4.1		3.8-5.6 mEq/L
Chloride	110		102-117 mEq/L

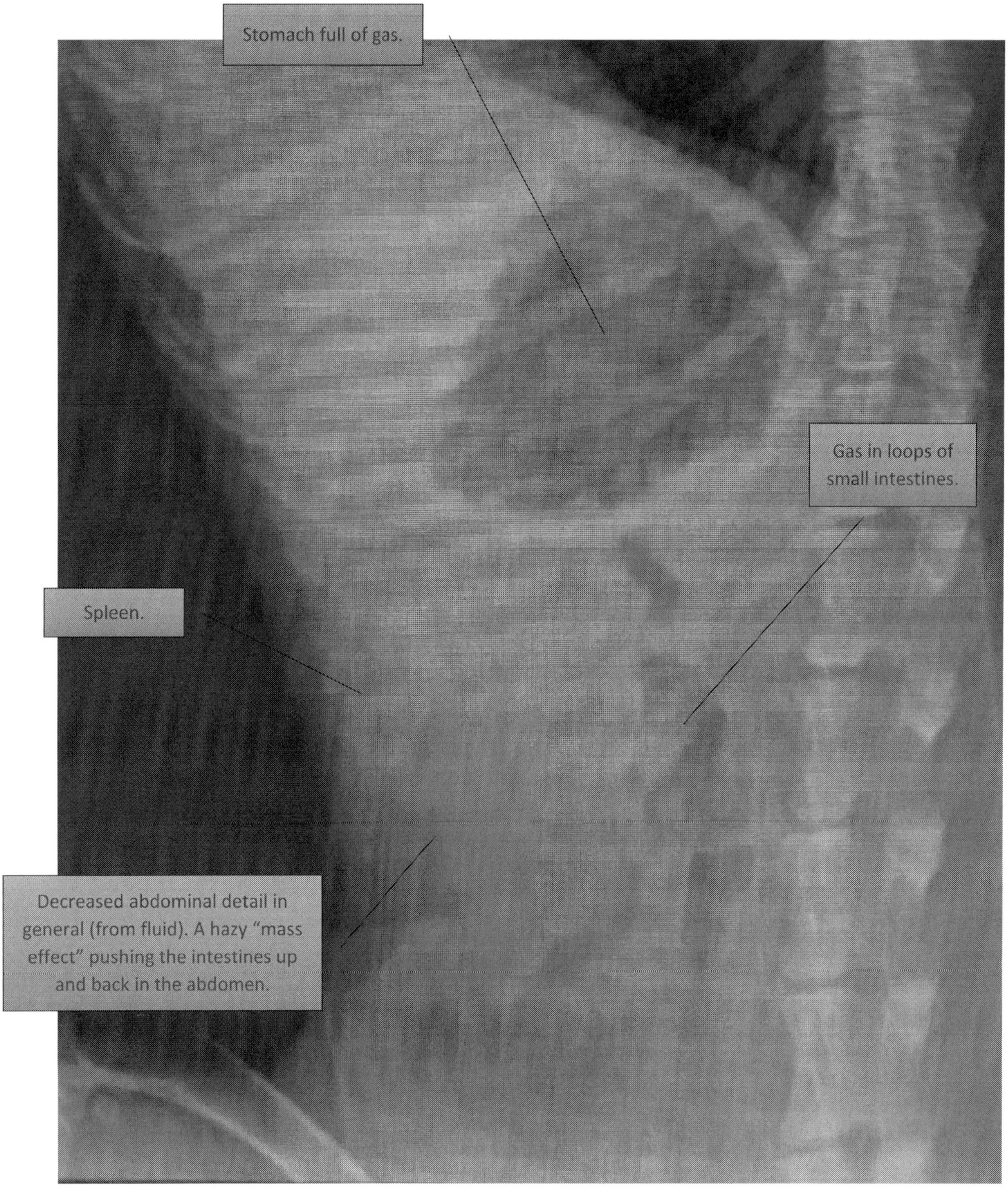

Right lateral abdominal view of "Flower" Melendez

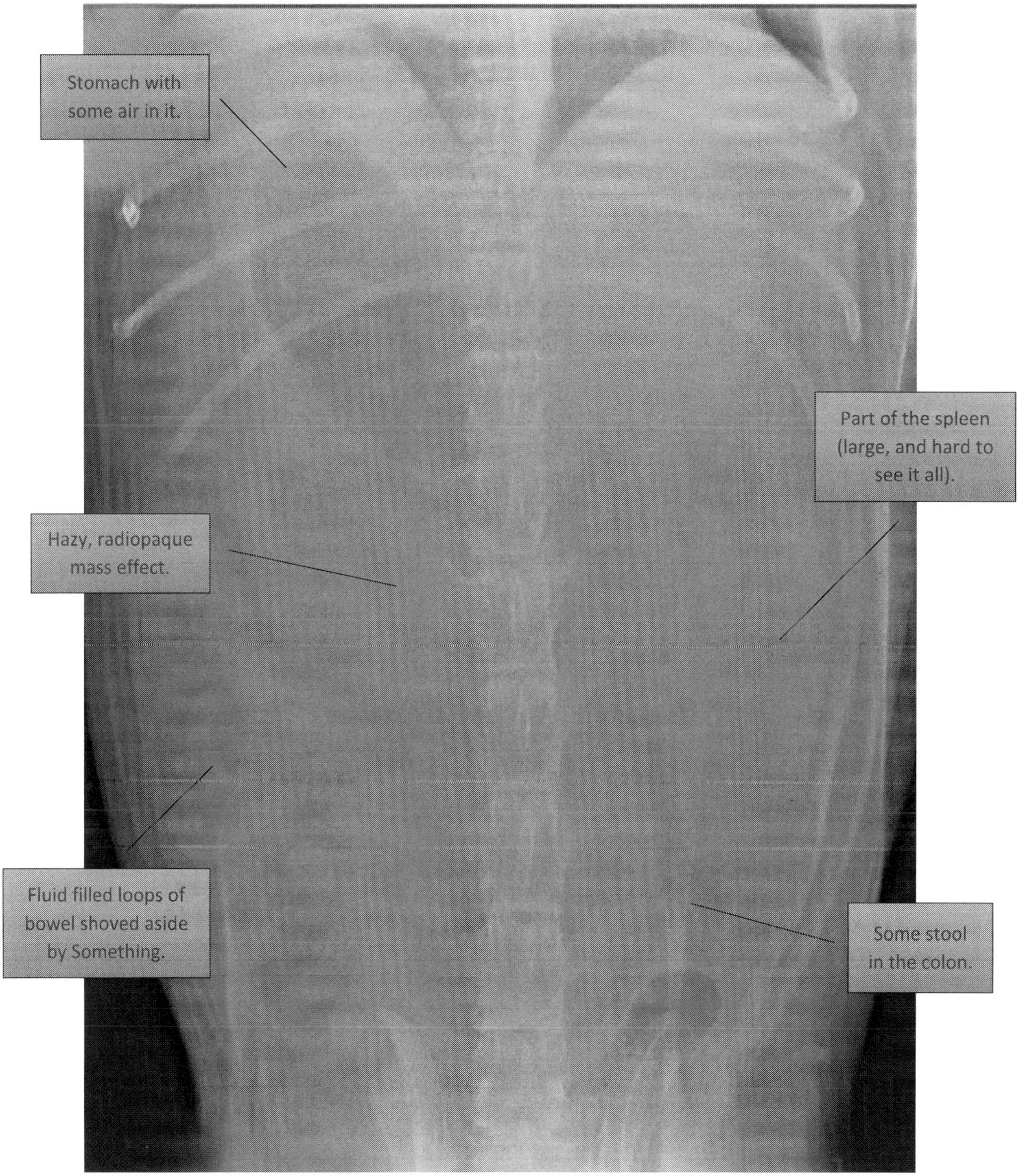

VD abdominal view of "Flower" Melendez

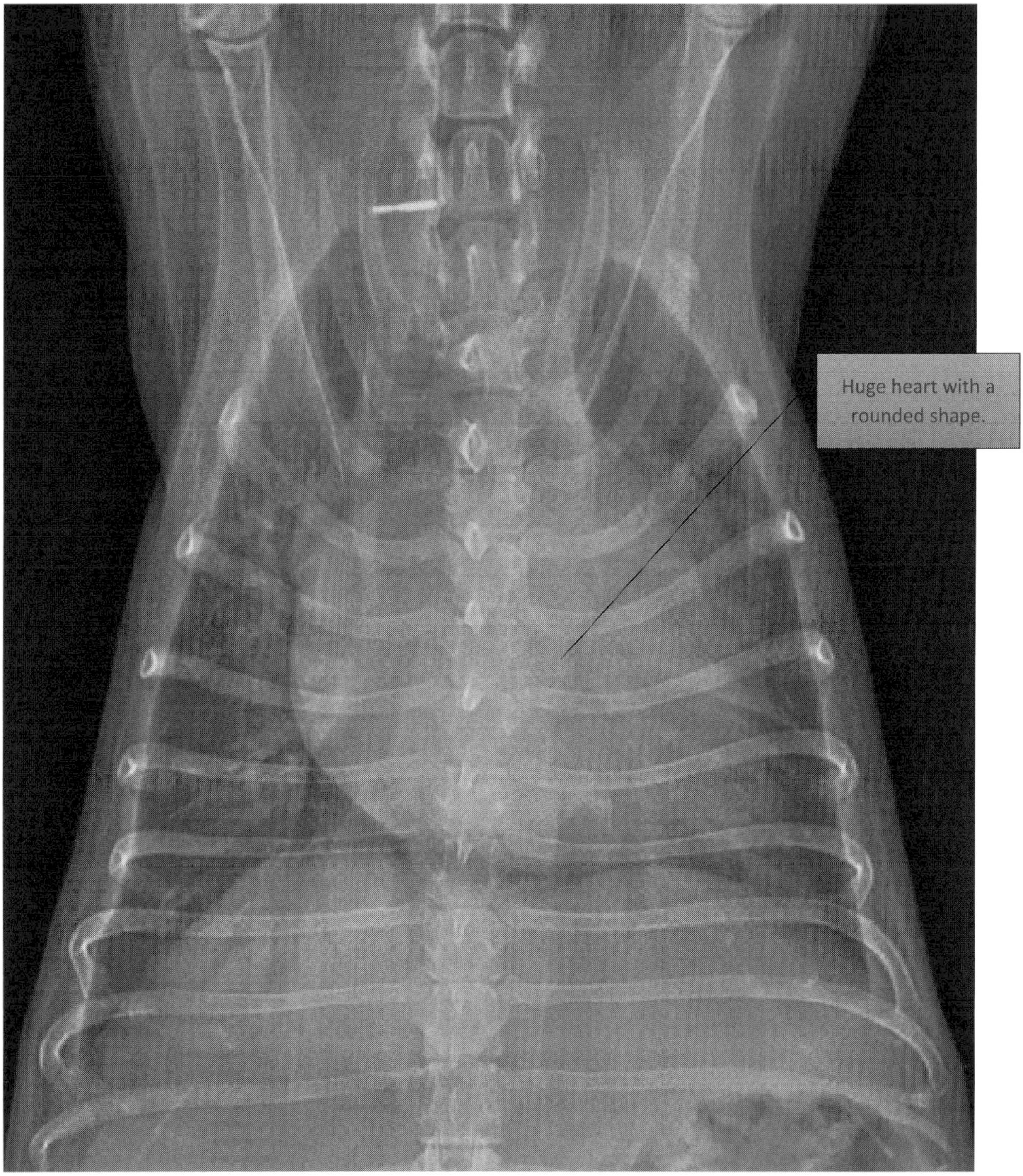

VD thoracic view of "Flower" Melendez

OK. Your patient is looking less and less stable, and you wisely decide to stop doing diagnostics so that she doesn't die on the radiograph table.

It's time to make some choices…

And your diagnosis is?

____

STOP!!!!!!

"Questions don't have to make sense, Vincent," said Miss Susan. "But answers do."

–Terry Pratchett, *Thief of Time*[xxxiv]

Poor Flower has **neoplasia** (cancer).

She has a splenic mass that's bleeding into her abdomen. The mass is visible on the spleen with your ultrasound probe. Though it's difficult to see the actual mass on the radiographs due to the fluid obscuring abdominal structures, you can see a "mass effect" that looks associated with the spleen, pushing loops of bowel out of the way. You can't be 100% certain that it's a tumor without a biopsy and histopathology (which is sending tissue from the mass itself off to a lab for an expert pathologist to do fancy tests on). But playing the odds based on all your experience and training (i.e., bringing the logic of Baysean Induction into play) would certainly lead to the conclusion that this dog has cancer.

Cancer is a broad word for diseases involving tissues that grow abnormally because of mutations or changes in the normal life cycles of the cells that make them up. Usually, this means that the normal cycle of cell death (programmed into a cell's DNA) is interrupted, and the cell goes on multiplying again and again. This makes tumors: masses of abnormal tissue.

There are many types of cancers named and characterized by the type of tissues they arise from. Some of them are benign and some are malignant. Malignant cancer can spread locally (i.e., just grow bigger where it's at) or metastasize (i.e. pop up in other parts of the body by sending cells to other tissues). Doctors will often use both staging (tests to see how much cancer has metastasized) and grading (tests to see how likely a cancer is to spread locally) to help guess how long a patient with cancer has to live.

You can hear a heart murmur. The heart is very large on radiographs, and you can see pericardial edema (fluid in the sack around the heart) on the ultrasound. That gallbladder "halo" sign is something you'll often see with heart conditions causing pericardial edema (as well as anaphylaxis, as we discussed in the last case).

What causes pericardial edema? There's a differential list for that! It includes neoplasia, congestive heart failure, right atrial tears, certain types of infections, migrating foreign bodies, diaphragmatic hernia, or extreme hypoalbuminemia.

So, there are lots of possibilities, right? Weeeeell... Another clue to consider is that *in dogs*, around 75% of pericardial effusion cases are from cancer (in cats it's usually from congestive heart failure). So again, playing the odds (and thinking only of the sign of pericardial edema), we're most likely dealing with cancer

Benign: cancer that doesn't grow and/or spread much.
Cancer: abnormal proliferation (growing and dividing) of cells to make tissue grow in wrong ways.
Grading: a system for estimating how abnormal cancer cells are to judge how fast they will grow.
Malignant: aggressive cancer that grows and/or spreads a lot.
Metastasize: when cancer cells move from a tumor to grow new tumors in other parts of the body.
Staging: a system for characterizing where in the body cancer has spread.

here. Recall your ultrasound image of the heart, with those unusual, rounded objects inside the chamber. Those certainly look suspicious of tumors inside the heart as well.

Now combine all these signs and we have a sadly convincing picture of a very serious neoplastic disease in Flower. A tumor on the abdomen has been bleeding in her belly and has metastasized to the heart and caused pericardial effusion. The most common type of splenic tumor in dogs (occurring in one study in 88% of dogs with a splenic mass and hemoabdomen) is hemangiosarcoma. This is a particularly malignant type of cancer that tends to metastasize to the heart and lungs quickly. Metastasis to the heart would make this a stage III (out of III) hemangiosarcoma. That's bad. What are you going to tell these clients? What options do we have to treat poor Flower? You know what to do!

Pick your treatments from the list…

a) Abdominal exploration and splenectomy
   Surgery to remove the spleen.
b) Abdominocentesis
   Sucking fluid out of the abdomen with a needle.
c) Antibiotics
d) A blood transfusion
   Giving the patient blood from outside.
e) Chemotherapy
   Combos of drugs designed to kill cancer cells.
f) A chest and abdominal splint
g) Diuretics
h) Drugs to increase the blood pressure
i) Drugs to slow bleeding
j) Euthanasia
k) Heart medications
l) Hire a professional mourner
m) IV fluids
n) Nebulization
o) Oxygen supplementation
p) Pain medications
q) Pericardiocentesis
   A procedure where you schluck out fluid from the heart sack to try and reduce pressure on the heart.
r) SQ fluids

There are hard choices to be made. Here are three basic treatment options. 1. **a and e**; 2. **j**; 3. **d, i and p**.

1. A **splenectomy** with a referral for **chemotherapy** would be one route we could go (especially if the Melendez family wants to try everything possible). If we want to attempt this, Flower will need some significant medical stabilization first. She's in shock from blood loss and is seriously anemic and would probably die under anesthesia in her current state. We need to pick some more options to go along with surgery and chemo. Which ones from the list do you think we would need? I picked **d, h, m, o, p and q.** She'll need a **blood transfusion** and probably **pericardiocentesis**—to reduce pressure on the heart. Usually, we treat patients in shock with **IV fluids**. However, this will make her anemia worse, and in cases of splenic tumors, aggressive amounts of fluids can cause faster bleeding from the tumor as well. So, we may have to go slow with IV fluids and use **drugs to help raise blood pressure** along with transfusion to help treat shock and hopefully keep her alive long enough for surgery. She'll need **pain medications**, and probably **oxygen supplementation** in the midst of all this as well.

   But will all this really help poor Flower? Sometimes, we will do surgery to remove the splenic mass and send it off to a lab. But there is strong evidence of involvement in the heart already, and we can't surgically remove cancer from the heart. A majority of cases with splenic hemangiosarcoma have microscopic metastasis to distant sites like the heart and lungs by the time they bleed in the abdomen: so even if we didn't have obvious evidence of metastasis, we'd be concerned. Studies of the mean survival time (MST) of dogs have indicated that dogs treated with surgery alone for splenic hemangiosarcoma (not considering staging) have between 19-86 days left to live. That's not a lot gained for such an invasive procedure with potential complications. Dogs that have chemotherapy after a splenectomy surgery for hemangiosarcoma have a better prognosis, MST of 274 days has been reported. But of course, there are side effects of chemotherapy to be considered.
2. A realistic alternative to all of this is **humane euthanasia**. This would be the best way to prevent further suffering in this already unstable patient (who could die at any moment) and keep her from having to go through treatments that will ultimately not do her much good.
3. Medical management (i.e., treating without surgery) is not a great option for this poor beastie. But if surgery is out and they refuse euthanasia, it's what we've got left. Some **pain medications** and **medications to slow bleeding**, preferably after stabilizing the patient with a **blood transfusion**,

Mean Survival Time (MST): a measurement of the average time from diagnosis to death in a patient, often used for communication about prognosis in cancer patients.

> would be the best we could do. This would not remove the problem, and it can all come crashing back down at any point, so this would essentially be to give a little more time for the family to say goodbye to Flower before the end.

You tell all of this sad news to Hector and Susan and stand awkwardly for a few moments while they blow their noses and wipe away tears. This is a case where there aren't any good answers. You tell them honestly that you wouldn't put your own dog through surgery and chemotherapy with cancer already in her heart and an HCT at 19%. "Being realistic, I don't see us getting far trying to save her." They nod their heads in sorrowful understanding.

"We know what we have to do," Mrs. Melendez sobs. "It doesn't make it any easier." She's right.

They opt to euthanize Flower. By this time, your technicians have placed a catheter (and have flow by oxygen running just in case). You give them the time they need to say goodbye. Flower dies as peacefully as possible.

You give both clients a hug and tell them you're sorry for their loss. Mean as she was, Hector and Susan Melendez loved their dog. "You've given Flower the last gift you can: an end to her suffering. You did the right thing for her."

It's a line you've repeated more times than you'd care to think about at this point in your career. But that doesn't make it any less true. Your furry patients know they're suffering, but they don't know why. And that makes it their people's job to know when it's time to stop.

Mrs. Melendez nods. "Little Samuel will miss her. She was always his favorite."

This makes all three of you cry again.

"Ven al restaurante pronto, doctor,"[13] Hector says, as he shakes your hand to leave.

You nod. It's not only a kind invitation but a good idea. Green chiles make everything better.

[13] "Come by the restaurant soon, doctor."

You wearily slump into your office chair. You can sure use a break!

It isn't too long, however, before Fred finds you and raps his knuckles on the open doorframe of your office.

"Come on in," you invite. "What's up?"

There's a look of concern on his face that makes you uncomfortable.

"There's something bothering me, Doc… Something we need to talk about." He stares at you intently. "Some of our clients are hiding things."

You're pretty sure you already know what he's noticed, but you'd really rather ignore it right now. Ah, but not wanting to deal with something doesn't make it go away.

What is bothering Fred?

Have you figured it out? Try to answer the question before you turn the page. If you need it, there's a hint in Appendix B.

"There is no point in using the word 'impossible' to describe something that has clearly happened."

—Douglas Adams, *Dirk Gently's Holistic Detective Agency*[xxxv]

CASE #12

"I think you already know something hinky's going down here, Doc," Fred leans against the wall, cradling his cup of coffee. He's rarely nervous like this. And he doesn't often use the word hinky. Nobody does, really. This is going to be a serious conversation indeed.

"I know," you sigh. "I know. Today and yesterday have brought it all to my attention."

"It runs deeper than I imagined," he says cryptically. "I think we have to do something about it."

You stare at the floor, pondering events, organizing the history of a series of crimes you've discovered.

The History: you can piece it together! There are hints on the next page and answers further on.

Two odd similarities

1.
2.

Step 1 – Organize the History

Signalment

Animals involved

Humans involved

Presenting Complaints (clues):

You've noticed some clues in your recent cases that somebody(ies) may be up to something(s) illegal. Write them here before turning the page.

1. The office break-in
2.
3.
4.
5.

Try to fill in some spaces before you turn the page to see some hints.

Some Hints. Use these to figure out what you are missing on the previous page.

Two odd similarities

1. Two histories are too close to the same story to be a coincidence.
2. Some wounds tell a story.

Step 1.

Animals involved: 2 dogs.

Humans involved: you've met three of them.

Presenting complaints: oddities from the wound stories and their owners.

STOP!!!!!!

Don't turn the page until you're sure you've done your best to fill in the events and the history...

The two animals and odd similarities

1. Both Asesino and Flower were supposedly attacked by coyotes.
2. Both Asesino and Flower had multiple scars, implying a history of lots of old wounds.

Presenting complaints (clues):

1. The office break-in
2. The wound pattern on both the Melendez dogs (Flower and Asesino)
3. Asesino's inconsistent wounds
4. Aggression of Melendez dogs
5. Something's off at La Chuleta (Hector Melendez's restaurant)

Now can you gather information on all these clues to figure out what's going on?

Step 2 – Gather Clues

Physical Examination (examine each clue):

Examine these events and identify some important aspects of each that stand out. List out observations about each.

The office break-in

1.
2.

The scars on both the Melendez dogs (Flower and Asesino)

1.
2.

Asesino's inconsistent wounds

1.
2.

Aggression of Melendez dogs

1.
2.

Something's off at La Chuleta (Hector Melendez' restaurant)

1.
2.
3.

STOP!!!!!!

Don't turn the page for all the answers until you've tried your best to fill in the lists...

Step 2 Answers

The office break-in

1. The perpetrator knew how to avoid cameras.
2. The medical records had been accessed, and it seems like someone was attempting to alter them (as evidenced by tampering with administration settings, etc).

The scars on both the Melendez dogs (Flower and Asesino)

1. There are a lot of scars (implying lots of past wounds).
2. Arturo seemed uncomfortable talking about the scars and had a dubious story about coyotes attacking his dogs multiple times. Hector and Susan told you the exact same story about Flower, which seems a *bit too coincidental.*

Asesino's wounds

1. Wounds are in multiple stages of healing, meaning there are scars (really old wounds that have healed), infected wounds (wounds old enough to have become infected), and fresh wounds.
2. Wounds are mostly distributed on the front end (head, neck, shoulders).

Aggression of Melendez dogs

1. Both Flower and Asesino were aggressive.
2. Both Flower and Asesino had the same story behind their aggression.

La Chuleta (Hector Melendez's restaurant)

1. The restaurant is chronically short on customers.
2. The restaurant is thriving financially (enough to pay Felicita Garcia and her coworkers generous salaries, and for Arturo to use his dad's money to pay for Asesino's medical bills).
3. Arturo is sharing in the benefits of his father's thriving business (using cash from it to pay).

Differential Diagnoses (list of possible explanations/solutions):

List as many possible explanations for each problem you can think of and pay attention to patterns.

1. The office break-in
   - ✓ ______________________________
   - ✓ ______________________________
   - ✓ ______________________________

2. The scars on both the Melendez dogs (Flower and Asesino)
   - ✓ ______________________________
   - ✓ ______________________________
   - ✓ ______________________________
   - ✓ ______________________________
   - ✓ ______________________________

3. Asesino's wounds
   - ✓ ______________________________
   - ✓ ______________________________
   - ✓ ______________________________
   - ✓ ______________________________

4. Aggression of Melendez dogs
   - ✓ ______________________________
   - ✓ ______________________________
   - ✓ ______________________________
   - ✓ ______________________________
   - ✓ ______________________________

5. La Chuleta (Hector Melendez's restaurant)
   - ✓ ______________________________
   - ✓ ______________________________
   - ✓ ______________________________

STOP!!!!!

Differential Diagnoses (the possible explanations):

1. The office break-in
   - ✓ People experienced in crime (mafia, gangs, professional thief, etc.) are interested in your business. (Leave town! Now!)
   - ✓ One of your clients wants to erase something that might have been written down in their record to avoid something incriminating.
   - ✓ One of your clients wants to modify someone else's records.

2. The scars on both the Melendez dogs (Flower and Asesino)
   - ✓ They are from random dog fights and coyote attacks in the past, just like Arturo says.
   - ✓ Arturo abused Asesino and Flower (before she was adopted) and is lying about it.
   - ✓ Hector and Susan abused Flower and Asesino and are lying about it.
   - ✓ Somebody else abused Flower and Asesino and Arturo is lying to protect them.
   - ✓ Both dogs have been in a dog fighting ring.

3. Asesino's wounds
   - ✓ They are from random dog fights and coyote attacks in the past, just like Arturo says.
   - ✓ Arturo abused Asesino before she was adopted and is lying about it.
   - ✓ Somebody else abused Asesino and Arturo is lying to protect them.
   - ✓ Asesino is part of a dog fighting ring.

4. Aggression of Melendez dogs
   - ✓ These are just mean ol' dogs, with bad personalities. Bad luck Melendez family!
   - ✓ Both dogs have learned aggression from bad luck: random fights with coyotes and other dogs.
   - ✓ Arturo abused Asesino (and Flower before she was adopted) and is lying about it.
   - ✓ Somebody else abused Flower and Asesino and Arturo is lying to protect them.
   - ✓ Both dogs have been in a dog fighting ring.

5. La Chuleta (Hector Melendez's restaurant)
   - ✓ Appearances are just deceiving: there are actually more customers than it looks like.
   - ✓ Someone's pumping money into the restaurant from an outside legal source.
   - ✓ Someone's pumping money into the restaurant from an outside illegal source.

OK… Can you pick one most likely differential from each list by assigning reasonable probabilities to items on the lists? Can you put together an accurate history based on the observations you've reached? Tell a story that might explain all these facts together…

Decision time!

Are there some tests you can run? Maybe try some mental ones…

- Are there recurring patterns running through these suspicious events?
- Is there a single explanation ("differential") that might explain more than one of these suspicious events?
- Which explanations seem most likely? Why?
- Who had the **motive, means,** and **opportunity** to carry out the crime of breaking into the clinic, and two other crimes you are now suspicious of?

And your diagnosis is?

STOP!!!!!!

The most likely differential from each list:

- ✓ Both dogs have been in a dog fighting ring.
- ✓ Asesino is part of a dog fighting ring.
- ✓ Both dogs have been in a dog fighting ring.
- ✓ Someone's pumping money into the restaurant from an outside illegal source.
- ✓ One of your clients wants to erase something that might have been written in the record to avoid some incrimination.

Can you put together a history based on the observations you've reached? What story would explain all of these facts?

You figured it out! The clues are all pointing toward the insidious workings of a **dog fighting ring**.

Asesino's wounds didn't make sense. He was covered in cuts and punctures *in different phases of wound healing* (scars, infected wounds, and fresh wounds). This must mean they were received on multiple different occasions, implying this dog is bit up all the time. In addition, Asesino's wounds were all on his front end, implying aggression: he's not turning to fight, but lunging forward even when wounded.

When Asesino's former housemate Flower showed up covered in older wounds this heightened your suspicions of organized dog fighting. She might not have been as aggressive a fighter (as shown by the clue that she has scars on her flanks from turning to run). That could be why Arturo didn't want her. It's possible someone could have even used her for "baiting" as part of another dog's training to fight (putting a timid dog out for a meaner dog to attack and learn aggression). Fighting dogs are trained in all kinds of cruel ways to be aggressive on purpose, which would certainly explain why both Flower and Asesino want to eat everyone.

Involvement in illegal dog fighting could explain why Arturo fumbled up a dubious explanation for his dog's wounds that was word-for-word the story he gave his parents about Flower (an exact repeat of the same story in a different dog could be a sign it was rehearsed).

You sigh again and sip some burned, lukewarm coffee. "Do you think we have enough evidence for a conviction?" you ask Fred.

"Probably not. But it's enough that we ought to report our suspicions to authorities."

"These are some of our best clients!"

Fred gives you a long look, under which you shrivel just a bit. "Does that really mean they should get away with abusing animals and breaking Federal laws?"

"No…but…what about Samuel?" Your words come out squeaky, and you try not to tear up as you think of the little boy dying of neurologic disease who sat in your office just yesterday.

Now it's Fred's turn to sigh. "Yeah…yeah, I know. There's no way to make that easier. If his father's really involved in this stuff, it's not Samuel's fault and it's awful that he'll have to deal with the consequences."

"I guess he can live with Hector and Susan."

"Wellllll…"

You look up sharply and read another topic in his eyes. "These dog fights generate a lot of money: illegal money. And illegal money is hard to spend if you don't want to get caught."

"So you need a way to launder it."

Fred nods. "You got it. I think the Melendez family businesses is a way to launder money for their son's illegal gambling and dog fighting."

"What exactly is money laundering?"

"Ah, well…" Fred goes into his geeky "lecture mode," a side of him you've seen before. Last time you fell asleep, but this time you feel more invested. "Money laundering is the shady process of converting money made from illegal enterprises (which could be traced back to the crime) and converting it into 'clean' money that's safe to spend without getting caught. It's usually accomplished by investing the dirty money in multiple legitimate businesses and fixing the books to look like the money came from them. The more investments you can make with the original cash, the harder it'll be to trace. I'll betcha Hector took the money from his son's criminal enterprises, injected it into his business, then hid it as income on his restaurant's balance sheets. Restaurant money could get pumped into real estate investments Susan Melendez brokers after that, to make it even harder to track. It usually happens in three steps, like this…"

He grabs a notebook on your desk and scribbles. You're almost as impressed with his quick sketch of a washing machine as you are with his mysterious knowledge of criminal activities.

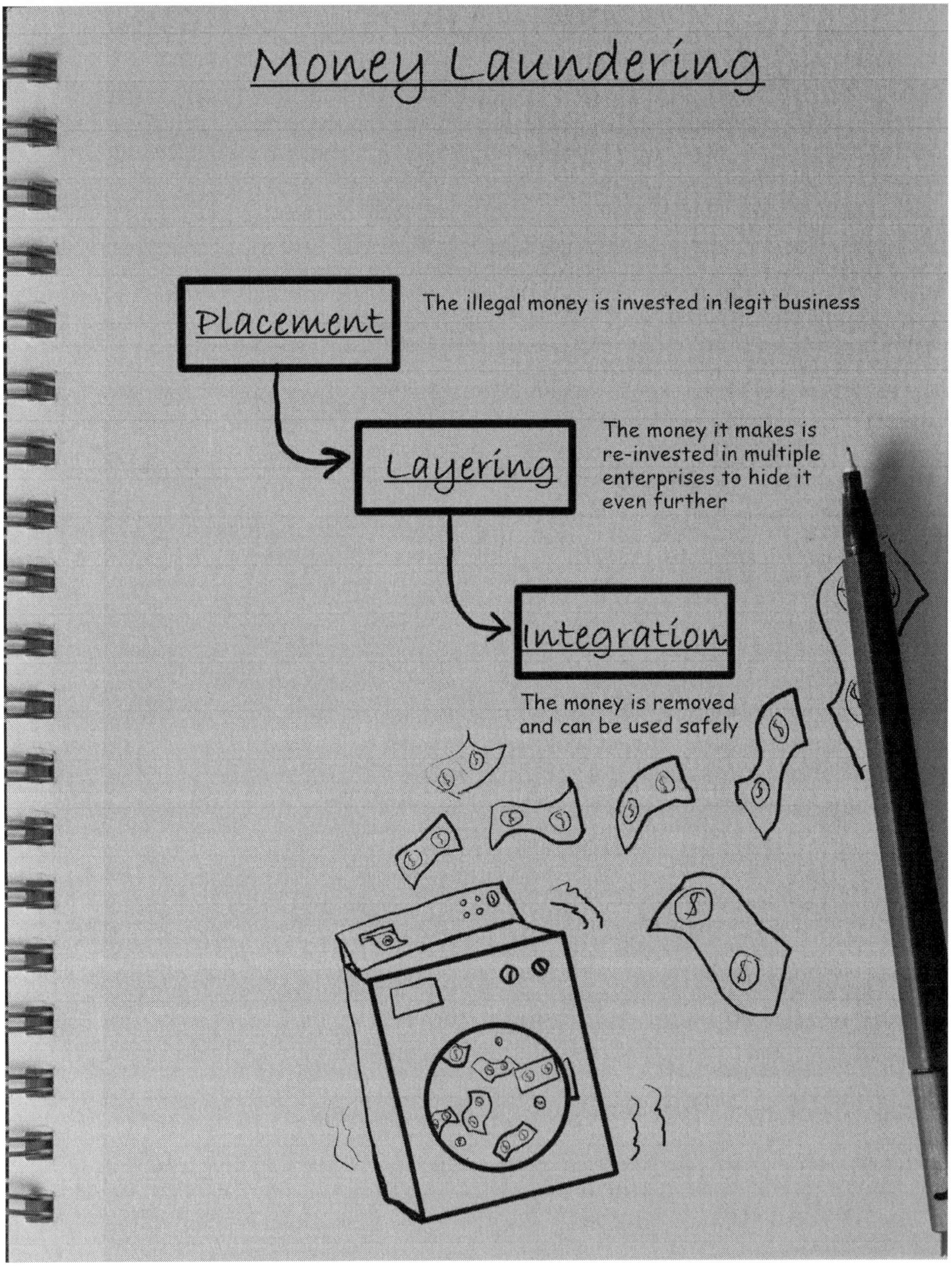
Money Laundering
Placement
The illegal money is invested in legit business
Layering
The money it makes is re-invested in multiple enterprises to hide it even further
Integration
The money is removed and can be used safely

"So..." the cogs in your brain connect as you speak. "That would explain why the restaurant does well, even though nobody but me seems to go there." Fred nods. "And why Arturo uses his dad's money (in large wads of cash) to pay for things."

Fred nods again. "Real estate investment is also a common way to launder money. The Feds may want to investigate Mrs. Melendez's business, too."

You run your hand through your tousled (yet so full and rich) head of hair. "This is a mess, Fred."

"It is. Maybe I'm wrong. Maybe it's all in my imagination. But we should tell someone," Fred replies.

"Yeah. I know a guy."

Fred grins mysteriously at you and leaves. You pick up your phone and call Agent Robert McCutchen from the CIA. He's a client of yours who drives his dog over from Albuquerque. He helped uncover a gang of counterfeiters among your clients last year. You have some really messed up clients, apparently. McCutchen thinks the situation is suspicious and says he'll call some friends to investigate.

You sit for a minute, pondering. You can't get Samuel's smiling face out of your mind. This could take his daddy away and ruin his chance of a cure. You scowl at the sketch he made on the wall, thinking how unfair it is that you're the one who has to make choices like this.

It's a sad, messed-up world: and sometimes doing what's right doesn't bring any warm, fuzzy feelings at all. Sometimes we have to do what we know we should despite what it will cost us, or even what it will cost people around us. Sometimes the reward for a good deed is misery.

Dog fighting is a terrible form of "blood sport" that involves pitting two dogs against each other to fight (sometimes to the death). The events are usually associated with gambling and other forms of illegal activity (like drug use), often run by organized crime syndicates for profit. It's illegal across all North America and is a felony in the US, prosecuted under the Animal Fighting Prohibition Enforcement Act of 2007. Since 2014 it's been illegal even to attend dog fights in the US.

Unfortunately, being illegal does not mean that dog fighting has gone away, and it's still something that occurs underground in every part of our country and across all kinds of subcultures. Some recent examples of dog fighting operations shut down included a large dog fighting ring with 30 people across five southern states, which the FBI (in cooperation with other organizations) shut down in 2009. "Operation Grand Champion" was a joint federal investigation that convicted multiple people across multiple states of dog fighting in 2019 (including a man from Albuquerque, NM with 30 years of experience in dog fighting) of involvement in dog fighting in 2019. 113 dogs were rescued in that operation.

Epilogue

The night is breezy, and the chill wind seems to blow even the shadows into undulating wraiths, as you make your way around the building after another long day at work. It's been two days since your phone call to McCutchen. You're the last one out again tonight since you had to wade through a pile of paperwork long after the staff went home. You gasp as a shadow suddenly detaches itself from the nearest corner of the building. A man in khaki pants and a flannel shirt bars your path.

"Hola, Doc." You recognize the voice. It's Arturo Melendez. You feel the hairs on the back of your neck stand up. "Working late again, eh?"

"Arturo!"

"Gotta follow up about your phone call to the Feds the other day."

"I...I'm sorry, Arturo. I had to. Dog fighting is a terrible thing, and I couldn't let it go."

"Yeah, well, you were right. Asesino and Flower were both fighting dogs. They brought in a lot of *dinero*, money for my boy...for Samuel. You understand, Doc?" He grabs you by the shirt collar as he chokes out the last words. You glance toward your car and try to estimate if you can beat him across the parking lot in a sprint. He's sure to have a knife, maybe a gun. He lets go and falls back a step. "You were right about my parents too... The money we made in the ring needed scrubbed, so Dad used his restaurant to launder the money. And Mom helped with a layer of real estate deals. It was a big operation, and the Feds came down on it like a hammer today. Like a hammer, *Gabacho*!" The unmistakable tone of rage and heartbreak mingle in his voice.

"I knew I should never have brought Asesino in and started worrying that night you'd figure out what really happened to him. That's why I broke into your clinic to see what you'd written in the records. But I couldn't change anything on that stupid program of yours."

"You broke in pretty easily." It probably isn't smart to say things like that right now, but you can't help your curiosity.

"Yeah. Been running with some bad *hombres* for a while. Had some practice with break-ins. At least I got some cash out of it for my boy. A man does everything he can for his family. But now..."

He takes a step toward you, and you get ready to run. He holds up a hand. "Wait! You were right about everything, Doc, except that Asesino and Flower were never my dogs!"

That stops you. "Wait... What?"

"They were my brother's dogs. My brother Raul."

"You have a brother?

"Yeah. Got a wife too. And a daughter in Washington, DC. She called last night worried about me."

"But you said Asesino was your…"

"I lied, man, I lied. Raul wasn't gonna bring Asesino in. He was just gonna let him die. But I knew. I mean, we're in the same gang. The Cobras run the fights on our land, but I never had anything to do with those. I wouldn't! But Raul was in them, he was in 'em big time. And I couldn't bring the policia down on my own brother, not on my own family. So, I brought Asesino, and pretended he was mine, and told you the same story I told Mom and Dad to give out about Flower."

"Well then, you're innocent! And you don't have to worry about the investigation."

He lets out a snort of disdain. "Innocent? The fights were always held on my land, and I took money Raul won from them for my son…lots of money. I wasn't brave enough to do the right thing and stand up against him. No. I'm not a good man. I'm in deep with the gang. I sold drugs. I broke into houses. I even mugged a guy. It was all to help Samuel, you understand. He needs those meds! And even if I was innocent, the Feds would arrest me for even being at a dog fight."

"I…I'm sorry Arturo." You can't think of anything else to say.

He looks at you. He sighs and puts and hand on your shoulder. "Samuel's why I'm here tonight. I'm going to jail. My parents and Raul are going to jail. I know this. My boy is going to need someone to watch out for him. Tell me you'll look out for Samuel."

You're stunned. This was the last thing you expected. "What? Me? Why me?"

"Because you're a good man. You do the right thing, even when it's hard. He needs that."

You stare at the darkness, your throat dry as you think about that dying boy. "I'll do what I can."

Arturo nods, a stern smile creasing his lips beneath streaks of tears along his face. He says nothing else to you but walks off into the darkness.

You call McCutchen, and he confirms warrants are out for Arturo, his parents, and his brother. You ask what's going to happen to Samuel. Apparently his mother's in drug rehab, and he'll be going into foster care. The next day, you start a funding campaign online to help pay for Samuel's meds. It doesn't matter that you and Fred Kawalkovitch will be the only two people to contribute to it over the next month. It's a step. It's something you can do.

And it inspires you to do more. You get busy pulling some strings. You're single and not ready to take in a kid, but you have friends who already foster with the state. Soon Samuel is living with them, and you check in on him regularly. His father begins the long process of conviction as the legal gears grind through his case. In the meantime, you'll do your best to get Samuel the medications he needs.

And one day, a brighter day while a dry New Mexico sun shines down on a grinning little boy's face (whom you've had the privilege of taking out for ice cream and park days twice a week all month): something beautiful has come out of all this tragedy. Doing the right thing may have felt abysmal, but in the end, your good deed did even more than save the lives of a couple of dozen fighting dogs: it's changed yours, and it's brought you an unlikely friend.

More Veterinary Adventures

If you paid close attention, you might say to yourself, "Wait! What about ______________?" Sorry! Some of the odd behaviors and suspicious events were red herrings.

Look for **FUZZY LOGIC ER**, the next installment in this riveting and mind-building experience, and see if you can find connections that run through 12 more amazing veterinary cases!

(scan for more Fuzzy Logics and Vet books)

If you enjoyed characters like Fred Kawalkovitch, Robert McCutchen, and Arturo Melendez you'll want to check out J. Aaron Gruben's time-bending fictional series **TANGLED EONS** where their adventures are more fully flushed out as they quest with veterinarians through time and more!

(scan to discover the Tangled Eons!)

Already read 'em? Please consider leaving a review on Amazon or other book sites, it helps a lot. Thanks!!

Disclaimer:

If and when you become a veterinarian, please do not attribute nefarious and criminal reasons to the odd behavior of your clients. Clients behave oddly all the time, and most of the time the reasons for it are perfectly innocent. I added a mystery to this book within the framework of the clients' histories just to make it more interesting (and to underscore the point that veterinarians and detectives both use inductive and deductive reasoning to solve cases). However, should you happen to solve a case or catch a criminal through these methods, I expect part of the reward money.

Glossary

Abdominal free fluid: abnormal fluid, floating around inside the abdominal cavity. Also called ascites.

Abdominocentesis: to schluck fluid out of the abdomen with a needle. A fluid tab of the abdomen.

Abscess: infection trapped under the skin, forming a pus bubble.

Acute: a condition that comes on suddenly.

AHDS: Acute Hemorrhagic Diarrhea Syndrome (sometimes called HGE, Hemorrhagic Gastroenteritis). This is a condition in dogs that makes them poop blood, while the blood is concentrated in their veins.

Albumin: a type of protein made in the liver and excreted in stool and urine.

Allergen: a harmless molecule that triggers the immune system into attacking it.

Alopecia: hair loss.

Anaphylaxis: a serious systemic or whole body reaction to an allergen.

Anechoic: An ultrasound term. A black area in the tissue, when no sound waves are bounced back.

Anesthetic: a drug that causes sleepiness (like for doing surgery) or numbness.

Antibiotic: a medication that kills or inhibits the growth of bacteria.

Antibody: molecules the immune system forms to fight allergens (and real infections).

Antiemetics: drugs that stop vomiting.

Antigen: a "foreign" molecule the immune system makes an antibody to.

Ascarids: the fancy name for roundworms and a broad group of intestinal parasites that attack all kinds of animals.

Auscultation: listening with a stethoscope.

BAR: Bright, alert, and responsive.

Benign: cancer that doesn't grow and/or spread much.

BID: twice a day.

Blood gas analysis: a blood panel specifically designed to check chemistry related to respiration and blood pH (like oxygen and bicarbonate levels).

Calculus (dental): a hard, porous coating over bacterial plaques on teeth. Synonym for tartar.

Cancer: abnormal proliferation (growing and dividing) of cells to make tissue grow in wrong ways.

Canid: a member of the dog species. (Canine is the adjective form of the noun canid.)

Capillary refill time CRT: capillary refill time. A simple way to check blood flow (perfusion) by pushing down on the gums, and counting how long it takes to pink up again in that spot. It should be less than 2 seconds in a healthy beast.

Cardiac Silhouette: the shape made by the heart on an X-ray.

Caudal vena cava: a big blood vessel that brings blood from the body back into the right side of the heart. It goes through the diaphragm and can be seen on ultrasound.

Caudal: positioned on the body closer to the tail than the head.

Chemistry panel: a blood panel that usually has basic information on the function of the liver, the kidneys, electrolyte levels, protein levels, and blood glucose.

Cholangeohepetitis: a disease where the bile duct and surrounding tissues in the liver and gallbladder are all inflamed.

Chronic: a condition that develops slowly over time.

Clotting times: a test of the ability of blood to clot properly.

Congenital: describing a condition a patient is born with.

Cranial: positioned on the body closer to the head than the tail.

Crepitation: a popping feel (in a joint's case, while you flex it). Can be a sign of arthritis in a joint.

Cyanotic: a lack of oxygen to the tissues, indicated by a purple color to the gums and tongue.

Cytology: looking at cells smeared on a slide in a microscope. If you pick this, you have to pick where to get the cells from. Will it be scrapings from the skin? Or cells from a tumor with a needle (fine needle aspirate)? Or cells from a swab of ear gunk or snot?

Cytoplasm: the "blood" of a cell—the central gooey stuff that the organelles float around in, surrounded by the membrane (lining) of the cell.

Differentials: the list of possible causes for a patient's problems. List of diseases that might make the patient sick.

Distention: gastric distention means the stomach is dilated and bloated.

Diuretics: medications that take fluid out of the body.

DMT plate: a special petri dish you can grow dermatophytes (ringworm) on.

Dorsum: the back, along the spine.

Dyspnea: abnormal breathing pattern.

Echocardiogram: ultrasound of the heart.

EOD: every other day.

Eosiniphils: a type of white blood cell that tends to increase with parasites or allergies.

Eupneic: a fancy word for breathing normally. The opposite of dyspneic (breathing abnormally).

Euthanasia: administration of a humane death, generally by injecting an overdose of an anesthetic agent IV.

Existentially quantified statement: a way to say that at least one of ______ exists out there.

Expiration: one of the two phases of breathing. You inspire when you breathe in, you expire when you breathe out. You also expire when you die, but that's a whole different synonym, baby!

Fecal direct saline exam: looking at a smear of poop in solution on a microscope (usually to examine the bacteria).

Fecal floatation: a test for parasites run by floating a gram of poop in a special solution for 10 minutes, then looking for parasite eggs under a microscope.

FeLV/FIV: feline leukemia virus and feline immunodeficiency virus. Two contagious diseases that suppress the immune system in cats.

Foreign body: a thing that's not where it should be, like something indigestible inside a loop of bowel.

Gastroenteritis: inflammation in the GI tract (often caused by bacteria or viruses).

GI: gastrointestinal. The stomach and intestines (supported by the spleen, liver, and pancreas).

Giardia: a microscopic parasite that looks like a flying saucer with googly eyes. It causes a nasty infection in the intestines that produces diarrhea and vomiting.

Globulin: a type of protein produced by the immune system to fight infection.

Grading: a system for estimating how abnormal cancer cells are to judge how fast they will grow.

Heart Murmur: a swishy sound on the stethoscope when listening to the heart. This is usually from a process that swishes the blood around (creates turbulent flow), like a problem with a heart valve.

Heartworms: worms that live in the right side of an animal's heart and cause heart failure.

Hemoabdomen: bleeding into the abdomen.

Hemoconcentration: elevated hematocrit (high red blood cells).

Hepatic: related to the liver.

Hepatomegaly: enlarged liver.

Hyperechoic: an ultrasound term. When an area on an image is seen as brighter than the surrounding tissue on the ultrasound machine (from more sound waves bouncing off tissue).

Hypersensitivity reaction: an unnecessary or overreactive immune system response to an allergen.

Hyperthyroidism: a disease process in which the thyroid gland produces too many hormones, affecting the body's systems.

Hypervolemia: too much fluid (in blood vessels in particular).

Hypocalcemia: low calcium levels.

Hypokalemia: low potassium levels.

Hypothalamus: gland in the back of the brain that controls hormones.

Ileus: sick intestines—moving too slowly.

IM: intramuscular. A shot into the muscles.

Inappetence: refusing to eat.

Infiltrative disease: a condition where cells invade places they shouldn't. Often we talk about this in the intestines, in particular, and often with types of cancer.

Inflammation: a response of the body's immune system to fight off disease, characterized by swelling and pain. It happens mostly through the mobilization of chemicals called inflammatory mediators and white blood cells.

Integument: a fancy word for skin.

Intussusception: an intestinal obstruction caused by a piece of intestine telescoping in on itself and getting stuck into a clump.

Ionized: having a charge on it (positive or negative molecules or atoms). *Not* the act of changing one's name to Ian.

IV fluids: fluids administered into a vein. The IV stands for intravenous fluids.

Lateral: looking from the side.

Leptospirosis: a bacterial disease that causes kidney and liver failure, often contracted through contaminated water (much like giardia). This bug looks like a coiled, swimming pigtail.

Lethargic: lacking energy and acting listless.

Luxation: being out of joint.

Lymph node: important structures for the immune system. "Hubs" of white blood cells and lymphatic drainage, to fight infection in specific places of the body. Small and bean-shaped, like the author of this book.

Macrophage: a type of white blood cell whose job is mainly to engulf and eat up invading bacteria and things.

Malignant: aggressive cancer that grows and/or spreads a lot.

Mammary glands: breasts. Mammary = produces milk, as mammals do.

Mange: mite infestation in pets.

Mean Survival Time (MST): a measurement of the average time from diagnosis to death in a patient, often used for communication about prognosis in cancer patients.

Metastasize: when cancer cells move from a tumor to grow new tumors in other parts of the body.

Modal logic: a branch of logic distinguishing modes of truth (like necessary vs. conditional truth).

Mucous membranes: the wet bits of a critter at the junction of skin and mucosa (like the gums in the mouth).

Nebulization treatment: Misting air moistened with saline (and sometimes a drug with it) over the nose to help with breathing

Neoplasia: cancer.

NSAID: non-steroidal anti-inflammatory medication. This is a type of pain medication that helps with inflammation, which is not a steroid. Aspirin is a common human one.

Nucleus: one of a cell's most important "organelles" (teeny organs), the nucleus is the brain of the cell and is where the DNA or RNA is stored. It stains darker than the cytoplasm on a cytology.

Nucleus:cytoplasm ratio: comparison of how bit the nucleus is to the amount of cytoplasm. This is one of many ways to help describe a cell under a microscope. A high N:C ratio = a bigger, fatter nucleus.

Nystagmus: a continual twitch of the eyes. Usually described as either horizontal, vertical, or rotary (round and round).

Oronasal: the mouth and nose.

Pancreas: an important little organ that makes enzymes and hormones to help digestion. He sits in the front of the abdomen, cheerfully tucked up against the first part of the small intestine near the liver and stomach. "Weird Al" Yankovic wrote a cool song about it.

Pancreatic Lipase: a test for pancreatitis. There's an fPL (feline pancreatic lipase) and a cPL (canine pancreatic lipase).

Pancreatitis: inflammation in the pancreas, which is an organ that makes hormones and helps digestion.

Parathyroid hormone: a hormone (made in a gland called the parathyroid gland, which is in the neck near the thyroid gland), which regulates calcium.

Parvo test: a test for canine parvovirus, which is a virus that attacks the GI tract in puppies.

PE: Physical Examination.

Pericardial edema: fluid building up in the sack around the heart (the pericardium).

Perilar: the area in the chest around the base of the heart.

Peripheral: outside of the brain or spinal cord. Then means limbs (when talking about musculoskeletal system).

Peritonitis: an infection inside the abdomen (outside the internal organs), causing inflammation of the peritoneum (the internal lining of the abdomen).

Petechiation: teeny little bruises on the skin or gums.

PMI: "Point of maximal intensity." The point where a heart murmur is the loudest.

Pneumomediastinum: air trapped in the mediastinum (the membrane "bag" that holds important structures in the center of the chest and divides the thorax).

Pneumothorax: air trapped between the lungs and the body wall. This is something often seen from punctures to the thorax.

Pocketing: in a wound, this is a term for an empty space formed by the skin pulling away from the body.

Polycythemia vera: a very rare disease where the blood is just super concentrated. This is one of the very rare cases we can still use medical leeches for! There are leech farms in the world for things like this…

Pregnancy test: a hormone test you can run on dogs at least 30 days pregnant.

Pre-renal azotemia: azotemia not from kidney failure.

Probiotic: a supplement of good bacteria that helps establish normal flora in the gut.

Ptyalism: excessive drooling. You could also say "hypersalivation," but it's not nearly as fun.

Pulmonary contusions: bruises (and bleeding) in the lungs.

Pupillary light reflex (PLR): testing if the pupils constrict normally when a light is shone in the eyes.

Pyrexia: a fancy word for having a fever.

QAR: quiet, alert, and responsive.

Radiographs: X-rays. A way to use radiation to take pictures of an animal's insides.

Renal: related to kidneys.

Rickettsial diseases: some infectious diseases carried by ticks (lyme is one of them).

SID: once a day.

Signalment: the descriptive characteristics of the patient.

Skin turgor: the tension of the skin, how well it falls back down when you tent it upward. This is a way we check the hydration status of a patient.

Splinting: tensing of the abdomen because of pain or fear.

sPO2: "saturation of peripheral oxygen". A measure of how much oxygen is getting to tissues.

SQ fluids: fluids administered under the skin. They're absorbed slowly over several hours thereafter. The SQ stands for subcutaneous.

Staging: a system for characterizing where in the body cancer has spread.

Sternal: having to do with the side of the body near the sternum (the "breastbone"). Often we use this word to mean a dog can sit up enough to lie on her chest.

Steroids: a class of chemicals (hormones or sterols) that inhibit inflammation and enhance metabolism. There are steroids made in your body, and handy steroid drugs made in laboratories.

Subcutis: the layer of fatty tissue just under the skin (helps insulate the body and bring blood to the skin, among other things).

Subluxation: a partial luxation.

Tachycardia: a rapid heart rate.

Tachypnea: fast breathing rate.

TBI: traumatic brain injury

Thoracic: related to the chest (also called the thorax).

Topical: to put on the surface of the skin (as opposed to giving it as a pill or an injection).

Triage: a quick assessment to decide how critically sick an animal is, for ordering which patients need to be seen first.

Ultrasound: a way to use sound in ultrasonic frequencies to view pictures of an animal's insides.

Universally quantified statement: an "if… then" statement phrased like "all ascarids are worms."

Urinalysis: a test on urine, generally looking for infection or crystals from bladder stones.

Urogenital: urinary and genital organs.

VD: ventrodorsal; looking from the chest toward the spine (patient laying on back) to take an x-ray: the beam is going from ventral (his belly) to dorsal (his back).

VHS: stands for vertebral heart score.

Visceral: related to the soft organs in the abdomen.

WNL: Within Normal Limits.

Wood's test: shining a blacklight on the skin to look for ringworm (a fungal disease of skin). Most cases of ringworm glow.

Xyphoid: the tip of the sternum.

Zoonotic: diseases that can be transmitted from animals to humans.

Appendix A: Bloodwork Panel Details

Client:	Species:	Sex:
Patient:	Breed:	
Patient ID:		

When you look at bloodwork, run your eyes over this "flag" column first: this cues you into which results were high (above the reference range) and which were low (below the reference range). Sometimes lab results are right on the border of the reference range. This might just as important as a number outside the range.

Chemistry Panel

Test	Result	Flag	Reference Range	
ALT			5-60	IU
ALP			10-150	IU
GGT			0-10	IU
tBili			0-0.4	mg/dL
Glucose			60-125	mg/dL
Cholesterol			112-328	mg/dL
Total Protein			5.1-7.8	g/dL
Albumin			2.6-4.3	g/dL
Calcium			7.5-11.3	mg/dL
BUN			7-27	mg/dL
Creatinine			0.4-1.8	mg/dL
Phosphorus			2.1-6.3	mg/dL
Sodium			141-156	mEq/L
Potassium			4.0-5.6	mEq/L
Chloride			105-115	mEq/L

These are the normal ranges for a patient.

ALT, ALP, GGT, tBili: These values will go up with ***liver disease*** or dysfunction (and a few other diseases).

Glucose: Glucose is blood sugar

Cholesterol: A fat derivative, useful for cell walls and other stuff

Total Protein: All the proteins: albumin is one type of protein

Albumin: A protein made in the liver.

Calcium: An important mineral stored in bones

BUN, Creatinine, Phosphorus: ***Kidney values***: they go up when the kidneys are failing (though there's some other stuff that can make them go up as well).

Sodium, Potassium, Chloride: These are electrolytes.

Client:	Species:	Sex:
Patient:	Breed:	
Patient ID:		

CBC

TEST	RESULT	FLAG	REFERENCE RANGE	
WBC			6.0-17.0	10x3/mm^3
Neutrophils			3-11.5	10x3/mm^3
Monocytes			0.18-1.35	10x3/mm^3
Lymphocytes			1-4.8	10x3/mm^3
Eosinophils			1-1.25	10x3/mm^3
Hematocrit			37-55	%
MCV			60-77	fL
Platelet Count			200-500	10x3/mm^3

WBC: This is the count of white blood cells which defend the body from disease.

Neutrophils, Monocytes, Lymphocytes, Eosinophils: Different types of white blood cells.

Hematocrit: The hematocrit (similar to PCV) is the concentration of red cells in the blood.

MCV: Mean Corpuscular Volume is a measure of the size of red blood cells: in general it will go up when the body is making new red cells.

Platelet Count: Platelets help blood clot.

Appendix B: Answer Key

Top Secret: Eyes Only

Case #7:

Step 1 – Organize the History

Be thinking about your list of possible suspects. All the client files that had been viewed were from yesterday, so especially consider the people you saw yesterday who might have acted suspiciously.

- The professionalism of the crime could be a clue. Which of your suspects might have experience in professional crime, or might have contacts with professional criminals?
- Could the description of the criminal from the glimpse on the camera, scant though it is, rule out any of your suspects? One's that don't fit this description could probably be ruled out (unless, of course, they hired someone).

Step 2 – Gather Clues

Physical Examination (PE):

<u>The front door:</u> It takes practice to pick a lock. This is not our burglar's first rodeo.

<u>Cash box</u>: did they want the cash or the receipts? Or perhaps both? Which of yesterday's clients would need cash enough to steal it?

<u>Millicent (your computer)</u>: Is it significant that it's your computer and not somebody else's? This indicates that it was one of *your* clients looking for information. Is it significant that Mole Rat's file was up? Could the tampering with settings and permissions tabs be an indication that somebody was trying to modify records? Seems likely! Consider reasons someone would want to modify their pet's medical record.

Differential Diagnoses (for motive):

- ✓ Robbery (cash): seems unlikely: $200 is seldom worth the prison time you'd get for B&E
- ✓ Robbery (something they didn't find): possibly…
- ✓ Robbery (something they stole and we didn't notice yet): also unlikely, since your staff and 4 police officers would have noticed.
- ✓ Coverup (trying to modify a record): possibly… Were there suspicious clients yesterday who had things to cover up?
- ✓ Information (wanted a client's info): maybe…. Was somebody after a client's information yesterday?

- ✓ Information (wanted a patient's info): possibly… But what would a person need to modify a pet's information in their medical record for?
- ✓ Information (wanted your info): not likely… There is value to your DVM and drug license. But those are also posted on the wall, so it seems unnecessary to break in and access the computer for them.

Case #8:

An upper respiratory infection triggered Grizwalda's hypocalcemia. The clues are: sneezing (in the history), a runny nose (on exam), and elevated white blood cell counts and globulins (on the bloodwork). An infection made her eat less (because of fever, feeling yucky, and smelling her food less). That meant she couldn't keep up with the massive increase in calcium demand her body needed to produce groceries for 6 hungry pups. And voilá, she has ze eclampsia!

Case #12: What's bothering Fred?

As an astute human being studied in the art of noticing significant patterns, Fred has noticed something today about a patient that matches something about a patient yesterday. Each of these instances left him with some questions. But the unlikely coincidence of the two together leaves him suspicious of foul play!

Endnotes-Works Cited

[i] Khan, Cynthia M., Editor. *The Merck Veterinary Manual, 9th Edition.* Merck & Co. Inc, Whitehouse Station, NJ. 2005. p2584

[ii] Khan, Cynthia M., Editor. *The Merck Veterinary Manual, 9th Edition.* Merck & Co. Inc, Whitehouse Station, NJ. 2005. p2586

[iii] "Douglas Adams Quotes." Goodreads.com. 1 November 2022. www.goodreads.com/author/quotes/4.Douglas_Adams

[iv] "Arthur Conan Doyle Quotes." Goodreads.com. 1 November 2022. www.goodreads.com/author/quotes/2448.Arthur_Conan_Doyle

[v] Khan, Cynthia M., Editor. *The Merck Veterinary Manual, 9th Edition.* Merck & Co. Inc, Whitehouse Station, NJ. 2005. p2584

[vi] Khan, Cynthia M., Editor. *The Merck Veterinary Manual, 9th Edition.* Merck & Co. Inc, Whitehouse Station, NJ. 2005. P. 2586

[vii] "Rex Stout Quotes." NotableQuotes.com. 1 November 2022. www.notable-quotes.com/s/stout_rex.html

[viii] "P.G. Wodehouse Quotes." Goodreads.com. 1 November 2022. www.goodreads.com/author/quotes/7963.P_G_Wodehouse

[ix] Tilly, Larry P. and Francis Smith Jr. *Blackwell's Five-Minute Veterinary Consult: Canine and Feline*, 6th ed. John Wiley & Sons, Inc. 2016. p1416.

[x] Tilly, Larry P. and Francis Smith Jr. Blackwell's Five-Minute Veterinary Consult: Canine and Feline, 6th Ed. John Wiley & Sons, Inc. 2016. Pp. 1416-1417

[xi] "G.K. Chesterton Quotes." Goodreads.com. 1 November 2022. www.goodreads.com/author/quotes/7014283.G_K_Chesterton

[xii] Khan, Cynthia M., Editor. *The Merck Veterinary Manual, 9th Edition.* Merck & Co. Inc, Whitehouse Station, NJ. 2005. p2584

[xiii] Khan, Cynthia M., Editor. *The Merck Veterinary Manual, 9th Edition.* Merck & Co. Inc, Whitehouse Station, NJ. 2005. p2586

[xiv] "Agatha Christie Quotes." Goodreads.com. 1 November 2022. www.goodreads.com/author/quotes/123715.Agatha_Christie

[xv] "7 Steps of the Decision Making Process." Concordia St. Paul Online. 2017. https://online.csp.edu

[xvi] "Hercule Poirot Quotes." Goodreads.com. 1 November 2022. www.goodreads.com/quotes/tag/hercule-poirot

[xvii] Khan, Cynthia M., Editor. *The Merck Veterinary Manual, 9th Edition.* Merck & Co. Inc, Whitehouse Station, NJ. 2005. p2586

[xviii] "P.G. Wodehouse Quotes." Goodreads.com. 1 November 2022. www.goodreads.com/author/quotes/7963.P_G_Wodehouse

[xix] "George Orwell Quotes." Goodreads.com. 1 November 2022. www.goodreads.com/author/quotes/3706.George_Orwell

[xx] "Terry Pratchett Quotes." Goodreads.com. 1 November 2022. www.goodreads.com/author/quotes/1654.Terry_Pratchett

[xxi] Khan, Cynthia M., Editor. *The Merck Veterinary Manual, 9th Edition.* Merck & Co. Inc, Whitehouse Station, NJ. 2005. p2584

[xxii] Khan, Cynthia M., Editor. *The Merck Veterinary Manual, 9th Edition.* Merck & Co. Inc, Whitehouse Station, NJ. 2005. p2586

[xxiii] "Peter Kreeft Quotes." Goodreads.com. 1 November 2022. www.goodreads.com/author/quotes/4939.Peter_Kreeft

[xxiv] "Hercule Poirot Quotes." Goodreads.com. 1 November 2022. www.goodreads.com/quotes/tag/hercule-poirot

[xxv] Leacock, Stephen. "Nonsense Novels: Easyread Super Large 20pt Edition." 2009. p.6. Web. ReadHowYouWant.com

[xxvi] Khan, Cynthia M., Editor. *The Merck Veterinary Manual, 9th Edition.* Merck & Co. Inc, Whitehouse Station, NJ. 2005. p2584

[xxvii] Khan, Cynthia M., Editor. *The Merck Veterinary Manual, 9th Edition.* Merck & Co. Inc, Whitehouse Station, NJ. 2005. p2586

[xxviii] "G.K. Chesterton Quotes." Goodreads.com. 1 November 2022. www.goodreads.com/author/quotes/7014283.G_K_Chesterton

[xxix] "Margery Allingham Quotes." Goodreads.com. 1 November 2022. www.goodreads.com/author/quotes/30748.Margery_Allingham

[xxx] "Ben Shapiro Quotes. Goodreads.com. 1 November 2022. www.goodreads.com/author/quotes/255605.Ben_Shapiro

[xxxi] "George Hammond Quotes." Goodreads.com. 1 November 2022. www.goodreads.com/author/quotes/383839.George_Hammond

[xxxii] Khan, Cynthia M., Editor. *The Merck Veterinary Manual, 9th Edition.* Merck & Co. Inc, Whitehouse Station, NJ. 2005. p2584

[xxxiii] Khan, Cynthia M., Editor. *The Merck Veterinary Manual, 9th Edition.* Merck & Co. Inc, Whitehouse Station, NJ. 2005. p2586

[xxxiv] "Terry Pratchett Quotes." Goodreads.com. 1 November 2022. www.goodreads.com/author/quotes/1654.Terry_Pratchett

[xxxv] "Douglas Adams Quotes." Goodreads.com. 1 November 2022. www.goodreads.com/author/quotes/4.Douglas_Adams

Picture Credits

Every effort has been made to attribute correctly all the material reproduced in this book. Unless otherwise noted, the pictures are public domain. Special thanks to North Texas Emergency Pet Clinic and Southern Veterinary Partners for the radiographic and ultrasound images.

Case #1 (Roundworm anatomy). *C. Elegans Male.* By K. D. Schroeder, own work (originally in color), CC BY-SA 3.0. Public Domain. https://commons.wikimedia.org/wiki/File:C_elegans_male.svg
Case #3 (Cross-Section of a Typical Eukaryotic Cell). *Animal Cell.* By Kelvinsong, own work. Public Domain (originally in color). https://commons.wikimedia.org/wiki/File:Animal_Cell.svg
Case #10 (The structure of an IgE molecule). *IgE.* By Sari Sabban, own work (originally in color), CC BY-SA 3.0. https://commons.wikimedia.org/wiki/File:IgE.jpg

About the Author

J. Aaron Gruben grew up in the Southwest and currently lives in Texas with his wife and six children, where he works full time as an emergenct veterinarian. A veritable Renaissance man, he is the author of works in multiple genres, including historical fiction, humorous fantasy, alternate history, Christian living, and chivalry and Viking school studies. When he is not repairing sick animals or writing, Aaron enjoys reading, hiking, acting goofy with his kids, playing board games, drinking a plethora of coffee, and making music on a variety of instruments. Christ brought Aaron into his camp at the age of 9, and his relationship with his beloved Savior has been the center of his life ever since.

Follow him on Facebook (*facebook.com/authoraarongruben*) and his blog (*nmgrubens.com/blog*) to keep up with new projects and random thoughts.

If you enjoyed the book, please consider taking a moment to leave a review on Amazon, Goodreads, or other book sites. Thanks!

Made in United States
Orlando, FL
10 November 2026

90498878R00114